Innovation und Entrepreneurship

Series Editors

Nikolaus Franke, Institut für Entrepreneurship und Innovation, Wirtschafts Universität Wien, Wien, Austria

Dietmar Harhoff, Max-Planck-Institut, München, Germany

Joachim Henkel, Dr. Theo Schöller-Stiftungslehrstuhl, TU München, München, Germany

Carolin Häussler, Universität Passau, Passau, Germany

W0037100

Innovative Konzepte und unternehmerische Leistungen sind für Wohlstand und Fortschritt von entscheidender Bedeutung. Diese Schriftenreihe vereint wissenschaftliche Arbeiten zu diesem Themenbereich. Sie beschreiben substanzielle Erkenntnisse auf hohem methodischen Niveau.

Innovative concepts and entrepreneurial performance are crucial for prosperity and progress. This publication series brings together scientific contributions on these topics. They describe substantial findings at a high methodological level.

More information about this series at http://www.springer.com/series/12264

Michael Vetter

Acquisitions and Open Source Software Development

 Springer Gabler

Michael Vetter
TUM School of Management
Technical University of Munich
Munich, Germany

Diese Dissertation wurde am 29.01.2021 bei der Technischen Universität München eingereicht und durch die Fakultät für Wirtschaftswissenschaften am 15.03.2021 angenommen

ISSN 2627-1168   ISSN 2627-1184 (electronic)
Innovation und Entrepreneurship
ISBN 978-3-658-35083-3   ISBN 978-3-658-35084-0 (eBook)
https://doi.org/10.1007/978-3-658-35084-0

Responsible Editor: Marija Kojic
This Springer Gabler imprint is published by the registered company Springer Fachmedien Wiesbaden GmbH part of Springer Nature.
The registered company address is: Abraham-Lincoln-Str. 46, 65189 Wiesbaden, Germany

Für Rosi

Foreword

About 20 years ago, Open Source Software (OSS) made headlines as a new way of developing software that, contrary to accepted wisdom, yielded very good results despite the lack of a central authority. Today, OSS and the OSS way of software development are a common and accepted part of the software industry. Interestingly, just as startups in other segments of the ICT industry, OSS startups are frequently acquired by established firms. This phenomenon raises a number of questions: Why is it that incumbents acquire software firms whose product—the code they develop—is largely available publicly and for free? How attractive for acquirers are OSS startups compared to startups in the same segment of the software industry that pursue a purely proprietary business model, and how is this reflected in the likelihood and timing of acquisitions? What role does an acquirer's own engagement in OSS play for acquisitions? And given that the OSS community around a startup's OSS project is an important resource of the startup, yet not under the control of the firm: What is the effect of an acquisition on the target's OSS projects and the related communities?

Michael Vetter addresses these questions in his dissertation. He combines thorough qualitative research, based on a large number of in-depth interviews, with a detailed quantitative analysis of huge amounts of GitHub data. His findings advance our practical knowledge and theoretical understanding of commercial OSS development, of acquisitions, and of the management of external resources.

This dissertation is the result of intense and dedicated research, performed with skill, enthusiasm, and commitment. It was a pleasure to be part of this endeavor as Michael Vetter's dissertation advisor. The findings presented in this work provide important new insights to research and practice. I recommend it to scholars and practitioners alike.

Munich   Joachim Henkel
June 13th, 2021

Acknowledgements

I am deeply thankful for having been part of an inspiring and encouraging community throughout my entire PhD. Therefore, I would like to take this opportunity to thank everybody who contributed to making my PhD journey such a rewarding experience.

First and foremost, I would especially like to thank Joachim Henkel, my thesis advisor, for giving me the opportunity to work on such interesting topics in such an inspiring and encouraging environment. With his deep passion for research, wealth of experience, and integrity, he has provided more support than I could have asked for. I would like to extend my gratitude to Reiner Braun for chairing my dissertation committee and to Jens Förderer for his kind agreement to act as my second PhD examiner.

I'm deeply grateful to my co-author Henning Piezunka, not only for the highly valuable advice and guidance on my research, but also the opportunity to visit him at INSEAD at their Paris/Fontainebleau and Abu Dhabi campuses, two enriching experiences during my time as a PhD student.

I would like to warmly thank the team at the Dr. Theo Schöller-Stiftungslehrstuhl für Technologie- und Innovationsmanagement, for valuable input, encouraging comments, and a fun time at the chair. Special thanks goes to Daniel Obermeier, for the many thoughtful conversations on academic and personal topics, your input lifting my research to another level, and the challenging games on the beach volleyball court.

My research would not have been possible without the help of numerous interview partners, to whom I want to express my gratitude for their commitment. I particularly want to thank Georgios Gousios for giving me the opportunity to discuss my approach to utilizing GitHub data and providing access to additional data.

I also want to thank BCG for supporting me during my time as a PhD student in many ways.

My research would also not have been possible without the support of TUM students helping me with data collection and proofreading. Therefore, I want to thank Katharina Beck, Marten Brandt, Maximilian Gebhard, Tim Godejohann, Fabian Heine, Johannes Huber, Marc Klingen, Sven Leeger, Sebastian Mair, Romain Parracone, Berkay Sapkaci, Janes Steiner, Sabal Subedi, and Anh Ma Tuan. I also want to thank my friends Max Knauerhase, Patrick Körner, Felix Thielemann, and Stefanie Weghorn for proofreading my dissertation.

Finally, I would like to thank my family for their continuous support and encouragement. I particularly thank my parents, Marion and Walter, for their trust, confidence, and unconditional support to pursue my personal and academic goals.

Zusammenfassung

Die Teilnahme von Firmen an der Entwicklung von Open Source Software (OSS) nimmt stetig zu. Ein großer Teil der heutigen OSS-Projekte wird durch informelle Zusammenarbeit zwischen Firmen und einer Community von Programmierern entwickelt. Mit einer wachsenden Anzahl an Firmen, die in der Entwicklung von OSS tätig sind, kommt es zunehmend häufig zu Akquisitionen solcher Firmen. Trotz der wirtschaftlichen und praktischen Bedeutung von OSS wurde dieses Phänomen bisher nicht erforscht. Die vorliegende Dissertation untersucht diese Schnittstelle zwischen der Entwicklung von OSS und Akquisitionen.

In der ersten Studie untersuche ich qualitativ die Rolle der Teilnahme von Übernahmekandidaten an der Entwicklung von OSS im Entscheidungsprozess potentieller Käufer. Die Studie basiert auf 52 Interviews mit Stakeholdern, die an verschiedenen Akquisitionen von in der Entwicklung von OSS tätigen Firmen beteiligt waren. Ich identifiziere Akquisitionsmotive im Zusammenhang mit der Entwicklung von OSS sowie verschiedene Möglichkeiten, wie die Teilnahme von Firmen an der Entwicklung von OSS die Entscheidungsprozesse potentieller Käufer beeinflussen kann.

In der zweiten Studie untersuche ich quantitativ die Rolle der Teilnahme von Firmen an der Entwicklung von OSS für Übernahmeentscheidungen. Unter Anwendung verschiedener Regressionsmodelle zeigt sich, dass junge Firmen, die an der Entwicklung von OSS teilnehmen, mit einer höheren Akquisitionswahrscheinlichkeit assoziiert werden, als Firmen, welche sich ausschließlich auf proprietäre Softwareentwicklung konzentrieren. Des Weiteren sind Käufer, die selbst in der Entwicklung von OSS aktiv sind, mit jüngeren gekauften Unternehmen assoziiert, als Käufer, die nicht in der OSS-Entwicklung aktiv sind. Ich diskutiere mögliche Mechanismen hinter diesen Beobachtungen sowie Implikationen für Theorie und Praxis.

In der dritten Studie untersuche ich die Auswirkungen von Akquisitionen auf OSS-Entwicklung, wobei quantitative und qualitative Methoden kombiniert werden. Es zeigt sich, dass Akquisitionen negative Auswirkungen auf die Anzahl an Beiträgen zu OSS von Mitarbeitern der gekauften Unternehmen und deren Community-Mitglieder haben. Variationen zwischen den Akquisitionen lassen sich auf Unterschiede in den Möglichkeiten und der Neigung der Käufer zurückführen, Ressourcen von den OSS Projekten der gekauften Firmen abzuziehen. Auch hier diskutiere ich Implikationen für Theorie und Praxis.

Abstract

Participation of firms in Open Source Software (OSS) development is steadily increasing. In fact, a substantial part of OSS projects today are developed in informal collaboration between firms and a community of voluntary contributors. As more and more firms are active in OSS, acquisitions of firms active in OSS development occur increasingly often. Yet, despite the economic and practical importance of OSS, research has so far overlooked this phenomenon. Therefore, this dissertation aims to explore this intersection of OSS development and acquisitions.

In my first study, I qualitatively explore the role of an acquisition target's involvement in OSS development for its acquirer in the pre-acquisition phase. Based on 52 interviews with stakeholders involved in various acquisitions of firms active in OSS development, I identify different acquisition motives related to a target's involvement in OSS development as well as further ways how the firm's involvement in OSS development can influence an acquirer's decision-making processes.

In my second study, I quantitatively examine the role of a firm's involvement in OSS development in the pre-acquisition phase. Combining OSS development data from GitHub with acquisition data from Crunchbase and using survival, competing risk, and ordinary least squares regression models, I find that young firms active in OSS development are associated with a higher likelihood of getting acquired compared to firms focusing solely on proprietary software development. I also find acquirers themselves active in OSS development being associated with earlier acquisitions of targets active in OSS development than acquirers not active in OSS development. I discuss the potential mechanisms behind these findings and implications for theory and practice.

In my third study, I examine the consequences of acquisitions for related OSS projects using a mixed-methods approach combining quantitative analysis with insights from the interviews. I find that acquisitions have a negative effect on the number of contributions by employees of the target and other developers contributing to the same projects as the target. I also find an enormous variation across acquisitions that can be traced back to differences in an acquirer's ability and tendency to extract resources from the target. The qualitative data from the interviews provide further insights into the underlying mechanisms. I discuss implications for theory and practice.

Contents

1 Introduction 1
   1.1 Motivation 1
   1.2 Research Objective and Design 2
   1.3 Structure of the Dissertation 7

2 Background: OSS Development and Acquisitions 11
   2.1 Basic Principles of OSS Development and Theoretical
        Foundations 11
        2.1.1 Brief History and Definition of OSS 12
        2.1.2 OSS Communities 15
        2.1.3 Motivation of Voluntary OSS Developers 17
        2.1.4 The Role of Companies in OSS Development 20
        2.1.5 How OSS Development Works in Practice 26
        2.1.6 Summary 29
   2.2 Basic Principles of Acquisitions and Theoretical
        Foundations 30
        2.2.1 Definition and Characteristics 30
        2.2.2 Brief Overview of the Typical Acquisition
               Decision-making Process 32
        2.2.3 Brief Overview of Strategic Acquisition Motives ... 34
        2.2.4 Key Aspects of and Relevant Research
               on the Pre-acquisition Phase 37
        2.2.5 Key Aspects of and Relevant Research
               on the Post-acquisition Phase 43
        2.2.6 Summary 48

2.3   Intermediate Conclusion: Two Perspectives Bringing OSS
      and Acquisition Research Together   49
      2.3.1   OSS Development as an Input to the Strategic
              Decision-making Process before an Acquisition   49
      2.3.2   Evolvement of OSS Development Activities
              as an Outcome of Acquisitions   50

3   **Qualitative Study: How Does the Involvement of Firms in OSS
    Matter in the Pre-acquisition Phase?**   53
    3.1   Introduction and Motivation   53
    3.2   Method   55
          3.2.1   Qualitative Study Design   55
          3.2.2   Sampling of Acquisitions and Interview Candidates   56
          3.2.3   Data Collection and Analysis   59
    3.3   Results   61
          3.3.1   Strategic View: OSS-related Acquisition Motives   61
          3.3.2   Process View: Roles of OSS in the Process Leading
                  to the Acquisition   70
          3.3.3   The Role of Licenses in OSS Acquisitions   75
    3.4   Discussion and Conclusion   77
          3.4.1   Summary of Findings   77
          3.4.2   Contribution   78
          3.4.3   Managerial Implications, Limitations, and Outlook   80

4   **Quantitative Study: The Role of OSS for Likelihood
    and Timing of Acquisitions**   83
    4.1   Introduction and Motivation   83
    4.2   Theoretical Background   86
          4.2.1   Factors Influencing Target Selection, Acquisition
                  Likelihood, and Acquisition Timing   86
          4.2.2   Firms' Potential Benefits and Downsides
                  of Engaging in OSS   88
    4.3   Data and Method   91
          4.3.1   Main Data Sources: Crunchbase and GitHub   91
          4.3.2   Sampling of Firms and Identifying Their GitHub
                  Activity   94
          4.3.3   Variables   96
    4.4   Results 105
          4.4.1   Descriptive Results 105

       4.4.2   Exploring Acquisition Likelihood of OSS-active
            and Non-active Firms 107
       4.4.3   Exploring three characteristics of a firm's OSS
            activities and their influence on acquisition
            likelihood 116
       4.4.4   The Role of an Acquirer's OSS Engagement
            for Timing of Acquisitions 118
  4.5   Summary of Quantitative Results and Discussion
      of Mechanisms 122
  4.6   Discussion and Conclusion 126
      4.6.1   Summary of Findings 126
      4.6.2   Contribution 127
      4.6.3   Managerial Implications, Limitations, and Outlook 128
5 Mixed-methods Study: The Effect of Acquisitions on Open
Source Software Development 131
  5.1   Introduction 132
  5.2   Theoretical Background 135
      5.2.1   Motivation of Contributors to OSS 135
      5.2.2   Consequences of Acquisitions 136
  5.3   Data and Method 137
      5.3.1   Sample Construction 138
      5.3.2   Variables 140
      5.3.3   Quantitative Analysis 141
  5.4   Quantitative Results 142
      5.4.1   Descriptive Results 142
      5.4.2   Contribution Activity after Acquisitions—Main
            Effect 145
      5.4.3   Variation in the Change in Contribution Activity
            after Acquisitions—Moderating Effects 147
      5.4.4   Robustness Checks 152
  5.5   Qualitative Evidence 155
      5.5.1   Structural Integration 156
      5.5.2   Acquirer's OSS Activity 157
      5.5.3   Role of Licenses 158
      5.5.4   Mechanisms Linking Acquisition and Community
            Response 159
      5.5.5   Alternative Explanation: Acquisitions to Eliminate
            OSS Competition 160

    5.6   Discussion and Conclusion 160
          5.6.1   Summary of Findings 160
          5.6.2   Contribution 161
          5.6.3   Managerial Implications, Limitations, and Outlook 163

6 **Summary and Outlook** 165
    6.1   Findings and Contribution 165
    6.2   Practical Implications 168
    6.3   Limitations and Future Research 169

Bibliography 173

Abbreviations

AGPL	GNU Affero General Public License
ASL	Apache Software License
BSD	Berkeley Software Distribution license
C-level	Company leadership with "chief" officer titles
CEM	Coarsened Exact Matching
CEO	Chief Executive Officer
CPHM	Cox Proportional Hazard Model
CTO	Chief Technology Officer
CV	Curriculum Vitae
DiD	Difference-in-Differences
DV	Dependent variable
GPL	GNU General Public License
IP	Intellectual Property
IPO	Initial Public Offering
LGPL	GNU Lesser General Public License
Ln	Logarithmus naturalis
MPL	Mozilla Public License
OLS	Ordinary least squares
OSI	Open Source Initiative
OSS	Open Source Software
R&D	Research and development
S.D.	Standard deviation
SIC	Standard Industrial Classification
VCs	Venture Capital firms

List of Figures

Figure 2.1   The GitHub workflow 27
Figure 2.2   The typical acquisition decision-making process 32
Figure 3.1   Acquisition motive coding and proposed extension
              of known acquisition motives with OSS-related motives 62
Figure 3.2   OSS-related acquisition motives and target's
              and acquirer's level of OSS activity 68
Figure 3.3   Information on target's OSS activities and risk
              reduction potential 72
Figure 4.1   Data sources utilized for GitHub data 93
Figure 4.2   Kaplan-Meier cumulative incidence rate (main model) 115
Figure 5.1   Average monthly commits to project around acquisition
              date 147

List of Tables

Table 4.1   Sampling steps and sample size   96

Table 4.2   Overview of variables used in quantitative analysis   97

Table 4.3   Descriptive statistics and correlations for all firms at the end of the observation period (incl. alternative events) 108

Table 4.4   Descriptive statistics and correlations for firms active in OSS development at the end of the observation period .. 109

Table 4.5   Descriptive statistics and correlations for target and acquirer characteristics at the time of realized acquisitions 110

Table 4.6   Comparison of descriptive statistics of firms active in OSS development and firms not active 111

Table 4.7   Regression results CPHM comparing firms active in OSS with non-active firms 114

Table 4.8   Regression results of competing risk model 116

Table 4.9   Regression results CPHM with focus on characteristics of a firm's OSS activities 117

Table 4.10   OLS regression results examining the role of the acquirer 120

Table 5.1   Descriptive statistics & correlation coefficients—Treated 143

Table 5.2   Descriptive statistics & correlation coefficients—Control 144

Table 5.3   Number of contributions to a project before and after acquisition: Main regressions 146

Table 5.4   Number of contributions to a project
             before and after acquisition: Moderators 149
Table 5.5   Number of contributions from the focal firm
             to a project before and after acquisition: Moderators 150
Table 5.6   Number of contributions from the community
             to a project before and after acquisition: Moderators 151
Table 5.7   Robustness check: Alternative specifications to test role
             of OSS activity of acquirer 153

Introduction

1

1.1 Motivation

Open Source Software (OSS) communities are a prominent example for user and developer communities producing technological innovation relevant for commercial applications (von Hippel and von Krogh, 2003). Even though the first OSS communities were born out of an ideological movement driven by higher motives behind the development of software such as freedom, independence, and equality (Raymond, 2001; Stallman, 1984), the importance of OSS communities to firms active in software development is clear (Fitzgerald, 2006). Firms have understood that developing OSS together with a community can be an external source for creativity and innovation, which can give them a distinct competitive advantage when utilized correctly (Dahlander and Magnusson, 2008). As a result, many firms today engage in and benefit from OSS projects, and informal collaboration between firms sponsoring OSS projects and voluntary contributors is an essential part of the OSS ecosystems (Bonaccorsi, Giannangeli, and Rossi, 2006; Colombo, Piva, and Rossi-Lamastra, 2014; Dahlander and Magnusson, 2005; Fisher, 2019; Macredie and Mijinyawa, 2011; O'Mahony and Bechky, 2008; O'Mahony and Lakhani, 2011; Perr, Appleyard, and Sullivan, 2010). In fact, a substantial part of the contributions to OSS stems from companies and their employees, and their involvement in OSS has been increasing steadily in recent years (Ho and Rai, 2017; Mehra, Dewan, and Freimer, 2011; O'Mahony and Bechky, 2008). Companies devote resources, including employee labor, to OSS projects, and in many cases companies provide much of the core development of those projects (Hann, Roberts, and Slaughter, 2013; Perr, Appleyard, and Sullivan, 2010; Spaeth, von Krogh, and He, 2015). Many highly successful OSS projects such as the Linux operating system, Firefox web browser, and the Apache webserver today benefit

© The Author(s), under exclusive license to Springer Fachmedien Wiesbaden GmbH, part of Springer Nature 2021
M. Vetter, *Acquisitions and Open Source Software Development*, Innovation und Entrepreneurship, https://doi.org/10.1007/978-3-658-35084-0_1

from such collaboration (West and Gallagher, 2006), and as a result, OSS has gained significant economic importance (Faraj, von Krogh, Monteiro, and Lakhani 2016; Nagle, 2019).

As more and more companies are active in OSS, acquisitions of firms active in OSS development occur increasingly often. Several high profile acquisitions like the acquisition of Sun Microsystems, the owner of the open source MySQL database management system, by Oracle in 2010, Zend, the main developer of PHP, by Rogue Wave in 2015, Xamarin, the main developer of the Mono framework, by Microsoft in 2016, or Red Hat, one of the biggest commercial contributors to many important OSS projects like the Linux kernel, by IBM for $34 billion highlight the growing importance of acquisitions in the area of OSS development. The increasing prevalence of this novel phenomenon raises several questions: For example, why do acquirers acquire companies whose software is most often available for free by downloading it from a development platform like GitHub or SourceForge? And how do acquisitions in which the companies contributing to OSS are involved affect the development of OSS? After all, evidence from prior acquisitions shows that outcomes of such acquisitions are rather diverse. For instance, the acquisition of Sun Microsystems by Oracle led to an outcry in the MySQL community, triggering several key employees to resign and sparked an ongoing dispute about the future openness of MySQL. In contrast, the acquisition of Xamarin, the key contributor to the Open Source cross-platform development tool Mono, by Microsoft was perceived positively by community members and employees. It resulted in the further open sourcing of proprietary Xamarin tools after the acquisition.

While the literature on OSS has recognized the role of companies in the development of OSS, this novel phenomenon of "OSS-acquisitions" has not been studied. However, understanding how the involvement of a firm in OSS development shapes acquisition behavior and acquisition outcomes and, vice versa, how acquisitions are shaping OSS development is particularly important given the proven economic value of OSS development (Faraj, Krogh, Monteiro, and Lakhani 2016; Nagle, 2019). The goal of this thesis is to start filling this gap.

1.2   Research Objective and Design

Both, research on OSS development and research on acquisitions are important streams of research in the management literature and have been extensively studied. However, the intersection of both has not been covered. Of course, just the lack of an intersection between two literature streams does not validate it as a field of study. This section shows how understanding acquisitions of firms active in OSS development can contribute to knowledge about OSS development and acquisitions. The section will also outline the key research objectives this thesis focuses on and the research design utilized.

The phenomenon of OSS development as a community-based and open innovation process has increasingly attracted scholarly attention across management disciplines. Extant literature covers the differences to proprietary innovation (e.g., Baldwin and von Hippel, 2011; Raymond, 2001), the motivation of individuals to participate in such communities (e.g., Hars and Ou, 2002; Hertel, Niedner, and Herrmann, 2003; Lakhani and von Hippel, 2003; Lerner and Tirole, 2002), the governance of OSS communities (e.g., He, Puranam, Shrestha, and von Krogh, 2020; Lerner and Tirole, 2005), and the evolution of projects and their communities (e.g., Fang and Neufeld, 2009; Ho and Rai, 2017; Shah, 2006; von Krogh, Haefliger, Spaeth, and Wallin, 2012). Increasingly, management literature has focused on the role of companies in OSS development. Literature and economic thinking have recognized that it can be beneficial for a firm to (selectively) open source code and participate in OSS development when done in an appropriate context at the right time (e.g., Dahlander 2005; Dahlander and Magnusson 2005, 2008; Gruber and Henkel 2006; Henkel 2006; Henkel, Schöberl, and Alexy, 2014). It has been understood that OSS development can be an external source of creativity and innovation to firms, which can give them a distinct competitive advantage (Dahlander and Magnusson, 2008). Openness in software development processes creates more opportunities for value creation—especially external value creation—compared with "closed" software development processes (e.g., Balka, Raasch, and Herstatt, 2014; Henkel, 2006; Jeppesen and Frederiksen, 2006), and profits from OSS development can be achieved if commercial pros of revealing code outweigh the cons (Henkel, 2006). Researchers have therefore studied the role of firms in OSS development from several angles: Their motivation to contribute (Fosfuri, Giarratana, and Luzzi, 2008; Henkel, 2004; Raymond, 2001; West, 2003; West and Gallagher, 2006), the business models they employ to benefit from OSS (Bonaccorsi et al., 2006; Chesbrough, 2006; Perr et al., 2010; West and Gallagher, 2006), the influence of OSS development on employees (Daniel, Maruping, Cataldo, and Herbsleb, 2018), and the relationship of firms and communities they engage with (Dahlander and Wallin, 2006; Ho and Rai, 2017; Spaeth et al., 2015).

Yet, OSS literature has so far not examined how the OSS activities of a firm—or more generally the openness of a firm—influences a firm's attractiveness for potential acquirers and their target search, selection, and evaluation behavior. One might even wonder: why do companies active in OSS get acquired at all? After all, at least parts of the target's code are typically available for free on the web rendering an acquisition to get access to the company's codebase unnecessary. On the other hand, the fact that the development takes place in the open potentially gives new opportunities for acquirers to evaluate a target based on its OSS activities—something not possible in the same way for firms focusing on proprietary software development or other forms of proprietary innovation as those innovation processes and their output are typically kept secret or at least protected. Hence, one might ask how do activities of potential targets in OSS development influence target search, evaluation, and selection processes of acquirers.

Acquisition literature does so far not provide guidance to answer these questions. While general acquisition motives have been understood, and some target, market, or acquirer characteristics relevant for target search, selection, and evaluation have been studied (e.g., Cunningham, Ederer, and Ma, 2020; Grimpe and Hussinger, 2008; Ouimet and Zarutskie, 2012; Ransbotham and Mitra, 2010; Warner, Fairbank, and Steensma, 2006; Worek, De Massis, Wright, and Veider, 2018), prior literature has not examined why firms active in OSS get acquired and how their OSS activity, or openness in general, influences target search, evaluation, and selection processes. Therefore, the first research objective for this thesis is:

- **Research objective 1**: Create an understanding of the role of targets' OSS activities for their acquisition.

This research objective particularly entails the goal to understand the motives behind acquisitions of firms active in OSS development and to understand the role of their OSS activity for target search, selection, and evaluation processes. Addressing this research objective is valuable from a phenomenological and a theoretical perspective. First, and from a phenomenological perspective, it is valuable as more and more (particularly young) firms participate in OSS development and open source their software (Gruber and Henkel, 2006; Hepp, 2016). Many of their founders and/or investors might seek to sell the firm when it matures. Hence, understanding the impact of a firm's decision to participate in OSS development on its potential future acquisition can be crucial for decision-makers at such companies and can help to understand the commercial value of OSS development of firms. Second, and from a theoretical perspective, understanding the role of OSS development of firms for acquisitions can be valuable as it provides a window to learn about the role of openness in the markets for corporate control and the decision-making processes leading to acquisitions.

However, whereas OSS development of the target might only influence the acquisition and the process leading to the acquisition, OSS development itself can also be affected by acquisitions. In particular, an acquisition may affect contributions to OSS projects both by the target itself and by the community, hence, changing the trajectory of OSS projects after an acquisition. Again, extant research has not examined this question and does not provide sufficient guidance to predict the effect of such acquisitions. On the one hand, OSS literature has examined some factors influencing the survival and success of OSS projects such as identification with the community, fun, or project quality (Fang and Neufeld, 2009; Ho and Rai, 2017; Shah, 2006), yet, little is known about the effect of major, potentially highly disruptive changes (such as an acquisition) at the target on OSS projects. On the other hand, acquisition literature has examined how being acquired affects the target's behavior and performance (e.g., Anand and Singh, 1997; Haspeslagh and Jemison, 1991; Homburg and Bucerius, 2006; Larsson and Finkelstein, 1999), but it is unclear how its contributions to OSS projects would be affected. It is also not clear how the community responds to the acquisition of the target. While research has begun to examine the response of a company's stakeholders to an acquisition (Bettinazzi and Zollo, 2017; Hernandez and Menon, 2018; Kato and Schoenberg, 2014; Kim, 2019; Rogan, 2014; Rogan and Greve, 2015; Valentini, 2016), there is no indication on how the community collaborating with the target responds to the acquisition. Therefore, the second research objective for this thesis is:

- **Research objective 2:** Create an understanding of the impact of an acquisition on the evolvement of a firm's OSS activities, the OSS projects it works on, and the related communities.

Again, addressing this research objective is valuable from a phenomenological and a theoretical perspective. First, and from a phenomenological perspective, it is crucial to understand the consequences of acquisitions for the various stakeholders, i.e., acquirer, target, community, and users relying on the OSS supported or developed by companies that get acquired. But understanding the impact of acquisitions on stakeholders is also important for the acquirers: For example, the acquirer might seek access to a community the target is active in. If the community reacts negatively to an acquisition and is not successfully integrated, the acquisition might destroy the very asset the acquirer seeks to acquire, and the value of the target can reach rock-bottom. Despite the frequency of acquisitions of OSS-sponsoring firms—the website index.co[1] documents 100 acquisitions of companies with an OSS business model for the years 2012–2018, with deal values

---

[1] https://index.co/market/open-source/acquisitions last accessed 06/15/2020.

of up to \$34 billion for the acquisition of Red Hat by IBM 2018 (one of the ten largest acquisitions by deal value in 2018[2])—we do not know about the effect of these acquisitions, neither on the target itself nor on the community it engages with. Second, and from a theoretical perspective, understanding the effect of acquisitions in the field of OSS can provide an opportunity to advance research on OSS and acquisitions as the dynamics in place in a community-based innovation model, characterized by informal collaboration such as OSS development (Lin, 2006; Rolandsson, Bergquist, and Ljungberg 2011; von Hippel and von Krogh, 2003), are different and more complex than in the traditional in-house innovation model. For example, managing a community outside the firm's hierarchical control requires specific skills that an acquirer might not possess. The acquirer might have a different attitude towards open innovation or seek to get rid of an OSS competitor by acquiring the sponsoring firm in the style of a "killer acquisition" (Cunningham et al., 2020). Or the acquirer might want to redeploy the employed developers working on the OSS project to other tasks or projects in the style of an "acqui-hire" (Coyle and Polsky, 2013), or want to change the direction of the OSS project. On the positive side, an acquirer may bring a superior resource endowment to boost the OSS project. The external community might react to all those strategic moves, which might again influence the trajectory of an OSS project. Hence, examining the impact of acquisitions on OSS development can help to refine our understanding of factors influencing how such hybrid organizations of firms and communities evolve over time (Ho and Rai, 2017; Shah, 2006; von Krogh et al., 2012).

This thesis follows an empirical mixed-methods approach to address the above research objectives, combining qualitative and quantitative data collection and analysis (Creswell and Plano Clark, 2011; Edmondson and McManus, 2007; Jick, 1979). By utilizing a mixed-method approach, this thesis can benefit from the advantages of both, quantitative and qualitative research methods. Qualitative methods are particularly suitable to investigate a novel phenomenon, provide insights into relevant constructs and their relationships, and capture causal processes (Eisenhardt, 1989; Eisenhardt and Graebner, 2007; Miles and Huberman, 1994; Yin, 2003). It is especially beneficial for studying research questions of how and why (Eisenhardt, 1989). In the context at hand, for example, one might wonder, "why are firms active in OSS acquired?" or "why does the community react the way it does?" In turn, quantitative methods allow investigating relationships in a broader empirical setting, which improves empirical generalizability (Creswell,

---

[2] https://en.wikipedia.org/wiki/List_of_largest_mergers_and_acquisitions last accessed 06/18/2020.

2014; Snow and Thomas, 1994). In the context of this study, the quantitative approach, for example, provides insights on how much and in which direction contributions to OSS change after acquisitions. Further, qualitative results were used to support quantitative findings by helping to interpret, clarify, and illustrate those results (Miles and Huberman, 1994). Bringing the advantages of qualitative and quantitative methods together, a mixed-methods research approach promotes both insight and rigor of results (Edmondson and McManus, 2007; Mahoney and Goertz, 2006). This is particularly useful as the research approach utilized when working on this thesis' underlying research project is also of exploratory character: I address both research objectives examining this novel phenomenon using an abductive approach (e.g., Bamberger, 2018) as existing theory from OSS development and acquisition research does not allow clear theoretical predictions—either, as I will show in the respective chapters, because different streams of theories predict different outcomes, or because of a lack of theory for this novel phenomenon.

The qualitative parts of this project draw on data from 52 interviews I conducted with C-level executives, founders, and employees from acquirers and targets, as well as community members of targets. Interviewing acquirers, targets, and community members allows capturing all relevant aspects of acquisitions of firms active in OSS development. The quantitative parts of this study draw mostly on data collected from GitHub and Crunchbase. Crunchbase is used for firm and acquisition data, which I combine with GitHub data that is used to capture OSS development activities. For robustness checks and validation of the data from the main data sources, I also use Orbis, LinkedIn, and Google Patents data.

1.3   Structure of the Dissertation

This dissertation is structured as follows. **Chapter** 2 introduces the basic principles of OSS development and acquisitions and the theoretical foundations thereof. First, Section 2.1 provides the basic principles of OSS development from a theoretical and practical perspective. Second, Section 2.2 provides the basic principles of acquisitions and related theories. The chapter concludes by explaining how the aforementioned research objectives bring both research streams together (2.3).

Chapter 3 presents the qualitative study on the role OSS activities of targets play as an input to the strategic decision-making process leading to an acquisition. This chapter's main objective is to validate the relevance of OSS activities of targets for acquisition decision-making and then further elaborate on how they are relevant. After an introduction (3.1), the qualitative research design, including the methodology for the sampling of acquisitions and interview partners and the analysis of the interviews are discussed in Section 3.2. In the subsequent section, the results are presented and discussed along three clusters that emerged from the coding of the interviews (3.3). The chapter concludes with a summary of key findings, the theoretical contributions, managerial implications, as well as limitations, and an outlook into areas for future research (3.4).

Chapter 4 provides the first quantitative evidence that OSS activities of targets play a role for their probability of being acquired, and briefly touches on the topic of acquisition timing. After an introduction (4.1), the relevant literature related to acquisition likelihood, target selection, and timing is presented (4.2.1). Combining insights from acquisition literature with literature on potential benefits and drawbacks firms can derive from participating in OSS development, potentially counteracting mechanisms of how the probability of a firm to get acquired might be influenced by its activity in OSS development are highlighted (4.2.2). The quantitative approach utilized in this chapter is explained in Section 4.3, including the main data sources (4.3.1), the sampling approach (4.3.2), and the variables (4.3.3). Section 4.4 shows the quantitative results along the three focus areas of the analysis: Differences between firms active in OSS development and firms not active in OSS development, differences within the group of firms active in OSS development, and the role of an acquirer's prior involvement in OSS. I discuss the findings from the quantitative analysis in light of theory and insights from the interviews in Section 4.5. The chapter concludes with a summary of the key findings, the theoretical contributions, managerial implications, as well as limitations, and an outlook into potential areas for future research (4.6).

Chapter 5 then focuses on the second research objective around the impact of acquisitions on OSS development. The chapter opens with an introduction (5.1). Potentially diverging effects of acquisitions on contributions to OSS from targets and their communities based on key findings from acquisition- and OSS-literature are highlighted (5.2). After an explanation of the exploratory mixed-methods approach, which combines various Difference-in-Differences regressions and interview findings (5.3), the effect of acquisitions on OSS development activities in projects is shown quantitatively in Section 5.4. I highlight differences between the effect of acquisitions on contributions from the targets' employees and from the communities they engage with, as well as differences across three key moderators. Potential theoretical explanations for the quantitative findings are discussed in Section 5.5 utilizing key findings from the interviews. Section 5.6 concludes the chapter with a summary of the key findings, the theoretical contributions, managerial implications, as well as limitations, and an outlook into potential areas for future research.

Chapter 6 concludes this dissertation with a review of the main findings and contributions for research and practice, key limitations to consider, and a summary of future research opportunities beyond the scope of this dissertation.

Background: OSS Development and Acquisitions

2

This thesis combines two important research streams—namely OSS and acquisitions research—that have not been connected previously. Therefore, the basic principles and foundations of both research streams are outlined in this chapter. It will first cover the relevant basic principles and theoretical foundations of OSS development and then explain relevant basic principles and theoretical foundations of acquisitions.

2.1 Basic Principles of OSS Development and Theoretical Foundations

OSS is defined as software under an OSS license (Raymond, 2001). While such software can be developed in-house or by a single developer, the term "OSS" also often implies that the software has been developed in the "OSS fashion" (e.g., Raymond, 2001; von Hippel and von Krogh, 2003)—that is, by an informal collaboration in public OSS communities (von Hippel and von Krogh, 2003; Lee and Cole, 2003). Such communities regularly consist of globally distributed participants that communicate over the internet (Crowston, Li, Wei, Eseryl, and Howison, 2007; Markus, Manville, and Agres, 2000; von Krogh and von Hippel, 2006). The phenomenon of OSS development and its specificities as a community-based and open innovation process has increasingly attracted scholarly attention across management disciplines. In this section, I briefly touch upon the historical roots of OSS development and elaborate on its definition (Section 2.1.1).[1] I will

---

[1] This history is highly abbreviated and I do not offer a complete explanation of the origins of OSS in this thesis. For more a detailed history, see for example DiBona, Ockman, and Stone (1999), or Raymond (2001).

© The Author(s), under exclusive license to Springer Fachmedien Wiesbaden GmbH, part of Springer Nature 2021
M. Vetter, *Acquisitions and Open Source Software Development*, Innovation und Entrepreneurship, https://doi.org/10.1007/978-3-658-35084-0_2

explain the core principles of OSS communities and the differences of software development in such communities of informal collaboration compared to traditional proprietary software development (2.1.2). I will then explain the motivation of individuals to contribute to OSS projects (2.1.3), as well as the role of firms in OSS and their motivation to contribute to OSS (2.1.4). Understanding firms' involvement in OSS development is fundamental for this thesis as only firms can be the target of an acquisition. I will then describe how developing OSS is done in practice (2.1.5), which helps to understand the variables I use in my quantitative analyses. Lastly, I will briefly summarize the key takeaways from this section (2.1.6).

2.1.1 Brief History and Definition of OSS

Prior to the emergence and clear success of OSS, economic theorists have long thought that this openness in innovation processes such as software development would be undesirable as openness would inevitably lead to the destruction of the incentives to innovate, as others would have the possibility to free-ride on the outcome of the innovator's labour (see, e.g., Dam, 1995; Granstrand, 1999). It was established thinking that a strong protection of proprietary knowledge, the exclusion of others, and the strict management of access, use, and imitation are a necessary prerequisite to adequately appropriate value from innovations (Plant, 1934). Therefore, countries have long offered intellectual property (IP) rights grants which—to a certain extent and temporarily—give inventors monopoly control over their inventions (Machlup and Penrose, 1950). The rationale has been that economic losses for society incurred because of IP rights can be more than offset by economic benefits from increasing investments in innovation and increased disclosure of knowledge, which otherwise would be kept secret (Foray, 2004; Machlup and Penrose, 1950). Yet, research has found that firms and individuals regularly voluntarily "freely reveal" their innovations, i.e., they voluntarily give up on exclusive IP rights, and all interested parties are granted access to it (Baldwin and von Hippel, 2011; Harhoff, Henkel, and von Hippel, 2003).[2]

While larger attention in media and research to the phenomenon of free revealing in OSS development has only been there since the early 2000s, the basic behaviors are much older in their origins and can be traced back until the late

---

[2] According to O'Mahony (2003) IP rights grants still can be useful to protect the inventors from liability even if they freely reveal their inventions. They can furthermore prevent expropriation of the innovator's innovations by others.

1960s (Lerner and Tirole, 2002). Back then, key elements of operating systems and the Internet were developed in academic environments in the US and large corporate research facilities such as Bell Labs (Lerner and Tirole, 2002). In these years, it was common for developers, who typically had a great deal of autonomy in their organizations, to share source code across different organizations.[3] Particularly relevant was the distributed development of UNIX, a mainframe operating system developed by AT&T. AT&T licensed the UNIX source code for free or a nominal fee to the developers because AT&T was not allowed to exploit UNIX as a commercial product due to governmental regulation (Weber, 2004). As they barely charged the developers for UNIX licenses, AT&T did also not provide any service and support to the developers. With no support available, the developers formed communities in which they supported each other and discussed their challenges and changes to the software (Lerner and Tirole, 2002; Weber, 2004).

In 1984, with UNIX increasingly becoming locked in by proprietary vendors, who were not willing to share their UNIX product's source code anymore, Richard Stallman created GNU, a free operating system resembling UNIX.[4] A year later, in 1985, he also founded the Free Software Foundation intending to protect free access to software and its source code for all software developers (Moody, 2001, pp. 14–23). "Free" in Stallman's view was not about software being available "gratis." It was rather about software for which users of the software had access to its source code and had the option to modify the source code and distribute the code they had modified (DiBona, Ockerbloom, and Stone, 1999; Weber, 2004). To ensure this "freedom" of software, Stallman and the Free Software Foundation introduced a formal license that aimed to prohibit "bad citizens" to sell proprietary software based on the OSS.[5] As part of the GNU General Public License (GPL), the users had to make any derivative works subject to the same license (i.e., GPL) and were thus unable to mix the OSS with closed source software and keep the result closed source (Lerner and Tirole, 2005).[6] The GPL is known to be the first OSS license and the first license of the "copyleft"-type. A copyleft license is a license that "grants everyone the right to use, modify and distribute

---

[3] Developers write source code using languages such as C, Python, and Java. If software is available only in binary form—i.e., a sequence of 0 s and 1 s—as opposed to the source code form, it is difficult for developers to interpret or modify the program (Lerner and Tirole, 2002).

[4] GNU is a recursive acronym for "GNU's not UNIX!" (see https://en.wikipedia.org/wiki/GNU; last accessed 25.01.2021).

[5] See https://www.gnu.org/bulletins/bull5.html for the GNU Bulletin in which the copyleft license was first introduced (last accessed 24.01.2021).

[6] See https://www.gnu.org/licenses/old-licenses/gpl-1.0 for the original GPL 1.0 license text (last accessed 24.01.2021).

the program on the condition that the license also grants similar right over the modifications he has made (Mustonen, 2003)." Today, several copyleft licenses with some variations of the requirements towards users and developers exist, such as the GNU Lesser General Public Licenses (LGPL), the GNU Affero General Public License (AGPL), and the Mozilla Public License (MPL).

Many developers sympathized with Stallman's idea of source code availability for pragmatic reasons. Yet, they regularly did not agree with Stallman's fundamentalism (Raymond, 2001; Weber, 2004). Those developers regularly wanted to combine proprietary code with "free" code and keep the result proprietary (Weber, 2004). However, the GPL does allow to keep software that integrates with GPL-licensed software proprietary. Furthermore, "free software" was a term considered unattractive for firms whose participation in the development and distribution of the software was considered to be a key success factor by some key developers in the "free software" movement. Consequently, the term "Open Source" was coined in 1998 by those developers (DiBona et al., 1999; Perens, 1999). They also defined "Open Source Software" as the software under an OSS license and founded the Open Source Initiative (OSI) as a governing body for OSS licenses. To be considered an OSS license, a license must adhere to the definition of the OSI. The OSI defined ten criteria for software to be considered OSS:[7]

1. *Free Redistribution of the Software*
2. *(Availability of the) Source Code*
3. *(Possibility to Make) Derived Works*
4. *Integrity of The Author's Source Code*
5. *No Discrimination Against Persons or Groups*
6. *No Discrimination Against Field of Endeavor*
7. *Distribution of License*
8. *License Must Not be Specific to a Product*
9. *License Must Not Restrict Other Software*
10. *License Must Be Technology-Neutral*

Whereas GPL-licensed software or software using other copyleft licenses is considered both, "free software" and "Open Source Software," other OSS licenses did not grant as much "freedom" as to consider the software using those licenses "free software" (e.g., Alexy, 2009). For example, frequently used licenses such as the MIT License (MIT), the Apache Software License (ASL), or the Berkeley Software Distribution License (BSD) allow the source code to be integrated into

---

[7] https://opensource.org/osd (last accessed 18.12.2020).

proprietary software without the requirement to release this software as OSS as well. Such licenses are called "permissive" licenses.

With this increasing formalization of the OSS activities and the rapid diffusion of the Internet, activities in OSS by individuals and firms, as well as research on such activities, have dramatically increased starting in the early 2000s. Today, a large number of OSS projects exist producing software for a wide variety of purposes. As of this writing, GitHub, the currently largest platform for OSS development, hosted over 100 million software projects.[8]

2.1.2 OSS Communities

As pointed out in the previous section, software is considered OSS if it comes under an OSS license. Technically, users or firms may develop the software internally and then license it under an OSS license. However, the term "OSS" also often implies that the software has been developed in the "OSS fashion" (e.g., Raymond, 2001; von Hippel and von Krogh, 2003)—that is, by an informal collaboration in public OSS communities (von Hippel and von Krogh, 2003; Lee and Cole, 2003). This is also how the term "OSS" should be understood in this thesis.

Like other user and open innovation communities, OSS communities are a fundamentally different organizational model for innovation than the organization of proprietary software development in traditional firms (Scacchi, 2004; Senyard and Michlmayr 2004; Shah, 2006; Vixie, 1999). In traditional software development, a team of software developers who are not the users of their software develops the software (see, e.g., Jones, 2003; Weber, 2004). The typically static team follows a pre-defined software development process, such as a sequential waterfall model (see, e.g., Cusumano, MacCormack, Kemerer, and Crandall, 2003), and releases the software to the company's customers once development is finished (Raymond, 2001; Senyard and Michlmayr, 2004).

Contrary to this, OSS communities are open to contributions from everybody interested, and the users of the software are encouraged to become involved in the community to develop the OSS further (Raymond, 2001; Weber, 2004). This is possible because the source code is released, allowing anyone to access, review, and change the code freely as the project progresses (Senyard and Michlmayr, 2004). OSS communities consist of a few to thousands of developers who contribute to the OSS development by writing code, bug reporting, testing, and

---

[8] https://venturebeat.com/2018/11/08/github-passes-100-million-repositories (last accessed 12.06.2020).

maintaining of the software, often contributing their time and effort voluntarily (Scacchi, 2004; Tuomi, 2002). Much of these activities are conducted by geographically and organizationally distributed contributors from all over the world (Bonaccorsi and Rossi, 2003; Lee and Cole, 2003).

There are different types of contributors to OSS communities, such as individuals contributing to OSS in their free time (often referred to as "hobbyists") and employees of firms contributing on behalf of their firm in their working hours (Markus, Manville, and Agres, 2000). Those contributors can distribute their creations widely and virtually without cost via the Internet, and more people from around the globe can be involved in the development of the OSS than internal company boundaries would usually permit (Dahlander and Magnusson, 2008). The result is that a large number of developers can obtain, test, and change the software for themselves (von Krogh and von Hippel, 2006), and the software can be adapted to the heterogeneous needs of hobbyists or firms (Franke and von Hippel, 2003).

The collaborative nature of OSS communities comes with potential advantages of the developed OSS over proprietarily developed software. Raymond (2001) famously made the point "given enough eyeballs, all bugs are shallow" (also called "Linus's Law"). The rationale behind this phrase was that if developers can use and look at the source code, they will be able to find errors in the software and point them out to the original developer of the software or even correct them on their own. Also, the more developers use and look at the software, the more likely it is that those bugs are uncovered and corrected (see, e.g., Alexy, 2009). This way, the software quality, reliability, security, and performance of OSS can be higher than those of proprietarily developed software (Lakhani and von Hippel, 2003; Morgan and Finnegan, 2007).

But the organization of OSS development in communities also has consequences relevant for this research: Due to the informal nature of the collaboration in communities, leaders of OSS projects do not have formal, contractual authority over voluntary contributors (Lerner and Tirole, 2002), and have only limited ability to discipline them (Raymond, 2001; Weber, 2004). Contrary to commercial software development, developers self-select themselves into the OSS projects they participate in and only participate in those activities they are most interested in or which are most beneficial to them (Crowston et al., 2007; Mockus, Fielding, and Herbsleb, 2005; Raymond, 2001). They can also end their participation in a community anytime they want without prior notice. Therefore, to achieve the advantages of OSS over commercially developed software mentioned above, the success of an OSS project stands and falls with the attraction of voluntary contributors to the respective OSS projects, and the long-term health of OSS projects depends on the ongoing willingness of contributors to continue contributing, and the joining of new contributors.

2.1.3 Motivation of Voluntary OSS Developers

A significant body of research has extensively studied why individual volunteers and firms publish or join OSS projects. I briefly elaborate on volunteer contributors' motivation in this section before explaining the role of firms in OSS development, including their motivation to contribute in the next section.

Research has pointed out that despite the public nature of contributions to OSS, developers may derive private rewards from their voluntary contributions to OSS projects (e.g., Raymond, 2001; Lerner and Tirole, 2002; von Hippel and von Krogh, 2003). Those rewards can be of intrinsic or extrinsic nature, and OSS developers most often contribute to OSS for a mixture of different intrinsic and extrinsic reasons.

Intrinsic rewards
Research has found that many individuals contribute to OSS projects for various reasons other than financial rewards (Lakhani and von Hippel, 2003; Raymond, 2001). Many of those are of intrinsic nature. An action is performed to gain intrinsic reward if done "for its inherent satisfactions rather than for some separable consequences" (Ryan and Deci, 2000, p. 56). I.e., developers contribute to OSS for the fun or challenge of doing it, and external pressures or rewards are somewhat irrelevant. As intrinsic rewards that make developers participate in OSS projects, research has identified enjoyment and fun, ideology, community identification, and altruism.[9]

Enjoyment and fun is one of the most important motives for participating in OSS development (Hertel et al., 2003; Lakhani and Wolf, 2005; Raymond, 2001). Developers decide to participate in OSS projects where they can solve technical problems they perceive as challenging and experience fun and enjoyment as the result of overcoming the cognitive challenges.

Ideology as a motivation to contribute to communities is linked to OSS development ever since Stallman initiated the free software movement (Stallman, 1999). OSS developers often adopt the ideologic belief that knowledge should be

---

[9] The terms and descriptions for the intrinsic and extrinsic rewards utilized in this section are heavily influenced by Sojer (2010).

accessible to everyone and participate in OSS development to realize this vision (Raymond, 2001; Shah, 2006).

Community identification describes the feeling of belonging to a community (von Krogh et al., 2012). Developers act in the community's best interest if they identify with it (von Krogh et al., 2012). Contributors tend to form bonds with their peers in the community and enjoy interacting with them (Shriver, Nair, and Hofstetter, 2013; Zhang and Zhu, 2010). Several researchers, including Hars and Ou (2002) and Lakhani and Wolf (2005), empirically confirmed that community identification is one motive for contributors to contribute to OSS.

Altruism describes the behavior of individuals seeking to help others while not expecting remuneration or reciprocity (Krebs, 1970). For example, altruism was found to be a key motive for Linus Torvalds to open sourcing Linux. He explained: "it feels good to have done something that other people enjoy using."[10] Other researchers, such as Hars and Ou (2002) and Wu et al. (2007), empirically confirmed that altruism is one motive for contributors to contribute to OSS.

Extrinsic rewards
An action is done to gain extrinsic reward if it "is done in order to achieve some separable outcome" (Ryan and Deci, 2000, p. 60). I.e., developers contribute to OSS for its instrumental value, and fun or the enjoyment which they gain from developing and contributing OSS are somewhat irrelevant. Research has identified personal needs, learning, signaling, peer recognition, and reciprocity expectations as extrinsic rewards that make developers engage in OSS projects.

Fulfilling *personal needs* is a key extrinsic reward motivating contributors to contribute to OSS projects (DiBona et al., 1999; Lerner and Tirole, 2002). For example, developers using the OSS may develop a new feature for the OSS they need; to ensure that this feature can also be used with future releases of the OSS, they contribute the feature back to the OSS project so it will be included in future releases of the OSS. Developers may also contribute to OSS when seeking solutions to challenges they face in their programming where the community might be able to provide solutions for (Raymond, 2001).

The opportunity to *learn* new skills and improve skills required as a developer is another extrinsic motive for developers to participate in OSS development (Hars and Ou, 2002; Lakhani and Wolf, 2005). As contributors can choose projects and tasks that meet their learning needs, participating in OSS is especially suited for learning software development skills while participating in real projects (Hars and Ou, 2002). Additionally, OSS projects often entail a peer review process

---

[10] https://firstmonday.org/ojs/index.php/fm/article/view/583/504 (last accessed 12.12.2020).

in which developers receive feedback from other experienced developers in the community (von Krogh, Spaeth, and Lakhani, 2003). The skills the contributors learn by contributing to OSS projects may ultimately help them get better job opportunities, more fulfilling jobs, and higher salaries.

Not only may OSS development be a good platform to learn skills, but it can also be a platform to *signal* skills and capabilities to potential employers and business partners, and thus help contributors to OSS advance their careers (Bonaccorsi and Rossi, 2003; Lerner and Tirole, 2002; Hann, Roberts, and Slaughter, 2013; Raymond, 2001). OSS developers have an incentive to publish and contribute code of high quality, as the openness of OSS inherently provides potential employers the possibility to evaluate a developer's skills based on the code they produced. In line with this argument, research has shown that higher ranks in the Apache community are associated with higher salaries (Hann, Roberts, Slaughter, and Fielding, 2002). Signaling rewards from OSS also explain why voluntary OSS contributors preferably join highly visible OSS projects (Lerner and Tirole, 2002).

Peer recognition by the community members has been found to be a motivation for developers to participate in OSS projects (Lakhani and Wolf, 2005; Lerner and Tirole, 2002). An essential part of the OSS culture is to give credit to the developers making the contributions to OSS projects (Raymond, 2001), and highly active and reputed developers can achieve the status of "stars" in OSS development (Wang and Tambe, 2020). For example, the famous freelance full-time OSS developer Sindre Sorhus has more than 40,000 followers on GitHub and Twitter.[11]

Lastly, *reciprocity expectation* describes the expectation of OSS developers to be able to count on the support of their community at a later point in time in exchange for their contributions to the community (Bergquist and Ljungberg, 2001; Lakhani and Wolf, 2005; Raymond, 2001; Zeitlyn, 2003). OSS contributors are motivated to contribute to OSS as a result of norms of mutual aid (Himanen, 2001). They expect to receive valuable feedback and contributions from the OSS community when they need it; and vice versa contribute back to OSS communities after they have received help from other contributors.

Having introduced the various motives why individual volunteers contribute to OSS, I want to explain the role of firms and their motivation to contribute to OSS in the next section.

---

[11] See https://github.com/sindresorhus and https://twitter.com/sindresorhus (last accessed 21.12.2020); See https://gitstar-ranking.com/ for a list of most followed users, companies, and repositories (last accessed 06.01.2021)

2.1.4　The Role of Companies in OSS Development

Many firms today engage in and benefit from OSS projects (Bonaccorsi et al., 2006; Colombo et al., 2014; Macredie and Mijinyawa, 2011; Peer et al., 2010). In fact, a substantial part of the contributions to OSS stems from companies and their employees, and their involvement in OSS has been increasing steadily in recent years (Ho and Rai, 2017; Mehra et al., 2011; O'Mahony and Bechky, 2008). OSS development is experiencing a shift from OSS as a community of individual developers to OSS as a community with many commercial companies, particularly small to medium-sized ones (Ågerfalk and Fitzgerald, 2008). Companies devote resources, including employee labor, to OSS projects, and in many cases, companies provide much of the core development of those projects (Hann et al., 2013; Perr et al., 2010; Spaeth et al., 2015). In this section, I elaborate on how companies engage in OSS, their potential benefits and drawbacks from their involvement in OSS, and how they interact with the community.

How firms engage in OSS development

Firms can engage with OSS in different ways. Research differentiates three modes of interaction of firms with OSS:[12]

- *Using OSS.* A firm uses OSS in its processes or products. This includes using (components of) the OSS to build proprietary software and use it in a firm's commercial products given the OSS licenses permit them to do so.
- *Participating in existing OSS projects.* The company actively deploys developers to contribute to already existing OSS projects (Dahlander and Wallin, 2006; O'Mahony and Bechky, 2008). For example, contributors contribute modifications of a piece of OSS created by the company for their own use back to the OSS project.
- *Releasing proprietary software as OSS.* The company starts a new OSS project by opening up the source code of a product they developed and let voluntary external contributors participate. This way, the company creates new OSS and the respective OSS community from scratch (Ho and Rai, 2017; Teigland, Di Gangi, Flåten, Giovacchini, and Pastorino, 2014; West and O'Mahony, 2008). Famous examples include Netscape open-sourcing the web-browser Mozilla, and Sun Microsystems releasing an early version of OpenOffice (Benson, Müller-Prove, and Mzourek, 2004; West and O'Mahony, 2005).

---

[12] This categorization is heavily influenced by Alexy (2009).

OSS is regularly used in the software industry, but also by many other firms, and has often become an integral part of commercial software products and corporate software architectures (e.g., Alexy, 2009). Firms that merely use OSS are not in the focus of this thesis, as almost all firms in the software sector probably do so, and pure usage of OSS is difficult to track. This thesis instead focuses on firms that engage in the latter two types of interaction with OSS—firms that actively participate in OSS development and sponsor employees to contribute to OSS. It is a significant step from using OSS to contributing to OSS for a company, as it is regularly against the company's policies that do not permit IP to leave the company or only permit it under a licensing agreement. Hence, a firm's active engagement in OSS should be significantly more important for an acquirer in its evaluation of the firm as a potential target than the mere use of OSS by the firm. Furthermore, since such firms can influence the direction of OSS projects, their acquisitions should be more relevant for the trajectory of related OSS projects than acquisitions of firms merely using the OSS.

The level to which firms engage in OSS can vary. It ranges from bug fixing or adapting the software to the firm's needs, and contributing the modifications back to the OSS project, to building the whole business model around OSS projects the company initiates and sponsors (such as, e.g., Red Had or Elastic). Releasing prior proprietary software under an OSS license and co-developing the OSS projects together with the community is the most radical step towards an open innovation model (Alexy, 2009; Chesbrough, 2003; von Hippel and von Krogh, 2003). Doing so usually also implies a different way of organizing and executing software development compared to proprietary software development where organization of the development team, management of the project, and coding have to be aligned with the community, and boundaries between the firm and its community may be blurred or even be erased (Chesbrough, 2003; Grand, von Krogh, Leonard, and Swap, 2004; von Hippel and von Krogh, 2003; West and Gallagher, 2006).

The engagement of a firm in OSS is not equal for all software they utilize or produce and not stable over time. First, research has found that firms mix different strategies of engagement in OSS, and that the right mix is key to be commercially successful (Fosfuri et al., 2008). For example, in the case of embedded Linux selective free revealing of OSS by firms was described by Henkel (2006). Firms partition their code into open and closed modules. They then collaborate on the open modules in OSS and develop the closed modules proprietarily (Henkel, Baldwin, and Shih, 2013). Second, a company's choice of engagement in OSS is not static. Companies may change their level of engagement in OSS over time. They can join a project and then increase their presence in projects. However, they can also decrease their engagement or stop releasing their software with OSS licenses and revert back to a proprietary model of software development and distribution (Hepp, 2016).

Firms' potential benefits and downsides from their engagement in OSS

The level of engagement of firms in OSS is closely related to the benefits and potential downsides that come with the engagement of a firm in OSS. While they share some of the extrinsic rewards individuals can attain by participating in OSS development (see previous section), there are further reasons for firms to participate in OSS development and open source their own code.

Similar to the "fulfillment of personal needs"-reward that individuals can gain from participating in OSS, firms often participate in OSS as a means to an end. Using OSS may provide access to high-quality software, and contributing modifications back allows to influence the official version of the OSS so that later releases of the OSS will be of more value to the company. For example, contributing back bug-fixes and modifications helps the firm ensure that those bug-fixes and modifications will be included in future releases of the software, and the firm will not have to re-do the improvements again. Furthermore, typical software engineering challenges such as interoperability and compatibility will be less of a worry for the company's developers (Raymond, 2001). Additionally, the contributions made by the firm's developers may be further improved by others, or the contributions can help to improve the credibility of the OSS project and thereby help to influence standard setting (Harhoff et al., 2003).

While individuals may participate in OSS to showcase their talent to potential employers, firms may participate in OSS to find and recruit those talents (Ågerfalk and Fitzgerald, 2008; Henkel, 2004). Contributing to OSS can also help companies to signal their technical competence and attractiveness as an employer to skilled developers (Henkel, 2004). Furthermore, developing OSS is seen as more prestigious and more enjoyable than proprietary software development. Thus, the motivation of a firm's employees working on such projects can potentially increase, and results improve (Colombo et al., 2014; Lakhani and von Hippel, 2003; Lakhani and Wolf, 2005; Raymond, 2001). Additionally, the skills that a firm builds as part of its engagement in OSS can be a valuable capability in increasing absorptive capacity (Cohen and Levinthal, 1990; Nagle, 2018a).

The increased visibility of a company stemming from its participation in OSS development is not only relevant to attract valuable talent, but also new users and customers. Given the low (monetary) boundaries to use OSS, open sourcing a firm's software can help attract new users to the software and help with the diffusion of a firm's technology (West, 2003). Thus, open sourcing a firm's software can help to achieve greater growth (Appleyard and Chesbrough, 2017). Furthermore, appearing as a good OSS contributor can help a firm increase its overall reputation inside and outside the firm (Raymond, 2001).

If a firm successfully built an OSS community around its own OSS, the OSS community is a valuable external resource to the firm (Dahlander and Wallin, 2006). The external contributors can make improvements to the OSS and add—potentially highly innovative—additions to it, and it is possible that the software will develop faster than if developed proprietarily (Dalle and Jullien, 2003; Henkel, 2004; Lakhani and von Hippel, 2003). Furthermore, the feedback from the community is often direct input from users of the software, which is valuable to better tailor the OSS to user demands and increase customer satisfaction. As customers are allowed to make changes and additions to the OSS themselves, they are more likely to be fully committed to the software and engage in its further development (Goldman and Gabriel, 2005; Morrison, Roberts, and von Hippel, 2000; von Hippel, 2001). Ultimately, the development efforts taken over by the community members decrease the time the firm's developers will need to spend on other activities not related to the firm's core activities (Goldman and Gabriel, 2005; Lakhani and von Hippel, 2003; Shah, 2006).

Another potential benefit a firm can gain from open-sourcing proprietary code and participating in OSS is an increase in demand and sales for complementary services and products the firm offers (Andersen-Gott, Ghinea, and Bygstad, 2012). The fast user adoption and diffusion of OSS technology increase the firm's user base. Those users—particularly if they are commercial companies themselves—often require complementary products or services at some point. Red Hat is a good example of a company successfully selling complementary services around Linux, to which Red Hat is a key contributor.

Research has also found that being active in OSS development can help firms cut product development or sourcing costs (Gambardella and von Hippel, 2019; Dahlander and Magnusson, 2005), a factor particularly relevant for younger firms, which often lack access to a large pool of resources (Gruber and Henkel, 2006; Hepp, 2016).

While research has found that a firm's being active in the development of non-pecuniary (free) OSS is associated with significant positive value-added (Nagle, 2018a), being active in OSS is not without potential downsides for firms. Contributing to OSS projects always means giving up some IP, which potentially could have been sold otherwise. As a result, monetizing software that has been open sourced is difficult and requires creative business models to profit from OSS (Chesbrough and Appleyard, 2007; Dahlander and Magnusson, 2008; Perr et al., 2010; West and Gallagher, 2006). Also, giving up IP can potentially mean a loss of competitive advantage as competitors can easily work with ideas contained in a firm's OSS at little to no cost (Henkel, 2006; West and Gallagher, 2006). Furthermore, building a community and motivating the community to supply an ongoing stream of external innovations and feedback is a key challenge for firms active in OSS (Daniel et al., 2018; West and Gallagher, 2006). Releasing OSS does not automatically attract many outside contributors who will do significant parts of the work and do so without remuneration (e.g., Alexy, 2009). Building a community requires time and money, as the code must first be prepared in a way so that the community can interact with it. For example, the code needs to be modular and well documented, so that (potential) community members can grasp the nature of the code, and inappropriate or other confidential comments need to be removed (Hecker, 1999; MacCormack, Rusnak, and Baldwin, 2006). Once the software is open sourced, firms must walk a fine line when desiring to obtain financial benefits from their investments in the OSS (Schaarschmidt, Walsh, and von Kortzfleisch, 2015), as the OSS community will less likely engage in the development of the software if the firm is perceived as being primarily driven by profit motives (Stewart, Ammeter, and Maruping, 2006). A firm that open sourced their software might even lose control over its future development if the community decides to continue a fork of the project and excludes the firm from participating by not considering the firm's contribution to the original project for the forked project anymore (Raymond, 2001).

Interactions of firms and the community
To ensure that a firm can extract benefits from their OSS engagement in OSS while avoiding potential downsides such as low engagement of the community or losing control of a project by means of a fork, research has examined the requirements for successful interactions of firms and communities in OSS. Understanding how successful interaction between firms and communities can be facilitated is key for a firm's success in OSS, as communities do not per se exist to support firms: Community members can and do freely choose what projects and tasks to engage in, and firms do not have formal control—as, for example, an employment contract would provide—over community member's decisions (Shah and Nagle, 2020; Spaeth et al., 2015). Thus, it is not surprising that it is a key challenge for companies to motivate outsiders to supply contributions to their projects (West and Gallagher, 2006), and a company's success is strongly tied to the successful management of the community (Perr et al., 2010).

One important factor influencing the interaction of firm and community are the governance mechanisms firms establish in OSS projects and how they exert control rights. Shah (2006) found that community members are aware of who controls decision and property rights in OSS projects. The level of control exercised in the OSS project influences (potential) community members' choice of whether or not to participate in the OSS project and a concentrated control can deter participation (Shah, 2006). Furthermore, Shah (2006) found that firms should not employ restrictions on wider code use and distribution by other contributors to their projects. License choice is a key aspect of OSS project governance in that regard (He et al., 2020; Lerner and Tirole, 2005; Stewart et al., 2006), and "outrageous license terms" beyond what is considered "fair" can have a negative impact on community participation (Shah, 2006). Actions considered "unfair" or opportunistic by firms holding control rights in an OSS project can change the behavior of community members and might lead to a decreasing participation of community members. If the community's expectations regarding governance are not met, many contributors may choose to refrain from contributing the code they created back to the project, resulting in the firm and other contributors having less build on and work with (Shah, 2006). As a result, tight control and some activities that permit value appropriation by the firm can be detrimental to value creation by the community (Shah, 2006).

Not only OSS project governance mechanisms implemented by a firm open sourcing their software matter, but also how the firm is perceived by the community plays an important role for attracting contributions from the community. When firms sponsor OSS projects, they need to exhibit certain characteristics to attract contributions by volunteers (Dahlander and Magnusson, 2005). Successful open-sourcing requires trust, openness, credibility, tact, professionalism, and transparency (Ågerfalk and Fitzgerald, 2008; Dahlander and Magnusson, 2005): The company and community need to establish a trusted partnership of shared responsibility when developing OSS software. Community members must believe in the credibility of the firm (i.e., the OSS developers' perception of its expertise and trustworthiness of the firm) and its openness (i.e., the firm's mutual knowledge exchange with the community) (Spaeth et al., 2015). Only then will the community members identify with the firm-sponsored OSS project and will be willing to contribute their effort and time to it (Spaeth et al., 2015). The firm needs to act professionally and signal high quality of its OSS work (Ho and Rai, 2017). Lastly, the firm needs to be transparent about its OSS project-related decision making (Ågerfalk and Fitzgerald, 2008). To maintain their perception in the eyes of the community, a firm needs to adhere to the rules of conduct of the OSS community and continuously foster their relationship with the community (Dahlander and Magnusson, 2005). Ideally, a firm supports the community members beyond the OSS development itself, for example, by providing technical assistance or even patented knowledge (Alexy, George, and Salter, 2013).

Lastly, firms need to manage their own employees in order to be successful in their OSS efforts, as those employees are the vital link between the firm and their OSS communities (Henkel, 2009). Employees of firms who participate in OSS projects can be committed to both their company and the OSS community (Chan and Husted, 2010; Daniel et al., 2018; Henkel, 2009). Yet, this commitment can vary, and employees can, for example, favor the community's interests over a company's interests. A firm active in OSS must therefore create positive experiences for employees contributing to OSS projects while also keeping them committed to the company (Daniel et al., 2018). To do so successfully, firms might employ not only formal policies, but also informal mechanisms to create a culture suitable for firm-sponsored OSS development (Chan and Husted, 2010). They might also staff employees on OSS or proprietary projects based on the employees' preferences (Daniel et al., 2018).

2.1.5 How OSS Development Works in Practice

To conclude my introduction of the foundations of OSS development, I briefly cover key elements of how OSS development is done in practice and how the interaction between community members is facilitated by taking GitHub as an example. This is particularly relevant, as the quantitative analyses in this thesis heavily rely on OSS development activity data from GitHub, the largest platform for OSS development (Gousios, Vasilescu, Serebrenik, and Zaidman, 2014), and many variables I utilize require a basic understanding of how OSS development works on this platform.

GitHub is a platform that allows developers ("GitHub users") to host and share their OSS projects and contribute to the OSS projects of other developers. GitHub also provides mechanisms that facilitate interactions between community members to enable the ongoing development and dissemination of OSS projects (Sims and Woodard, 2020).

OSS projects on GitHub, similar to any other OSS development platform, are often initiated by an individual or a company that desires to develop software to fulfill their own needs (von Krogh and von Hippel, 2006). They typically develop an initial version of the software on their own, utilizing a traditional software development approach, i.e., without informal collaboration between distributed individuals (Senyard and Michlmayr, 2004). However, they need to prepare the project in a fashion so that the community can contribute to the OSS project. Therefore, the project should ideally offer interesting tasks and a critical mass of code with which the OSS community can interact (Lerner and Tirole, 2002;

Raymond, 2001; von Krogh et al., 2003). Once an initial version of the software is created, it can be published on the GitHub platform (github.com). Every GitHub user (i.e., the developers) can create public OSS projects there that other GitHub users can download, utilize, and contribute back to. Such OSS projects are also called "repositories," and the GitHub user creating the repository is called the "owner" of it.[13] The developers can also add documentation or license files to each project. Firms or other organizations can have their own GitHub accounts and can permit their employees to, for example, create and change projects in the organization's account.

Two important roles must be differentiated in the projects: "Maintainers" and "contributors." Maintainers have certain rights in a project and can make *direct contributions* to it—i.e., they can directly change the source code of the project. Owners of projects are also always maintainers, but owners can also assign maintainer rights to other GitHub users. Contributors can only make *indirect contributions* to a project (Vasilescu, Van Schuylenburg, Wulms, Serebrenik, and van Den Brand, 2014)—i.e., they can propose changes to the source code, and the maintainers are free to accept or reject the proposed changes (Sims and Woodard, 2020). This process of direct and indirect contributions is facilitated in GitHub by a set of complementary mechanisms depicted in Figure 2.1.

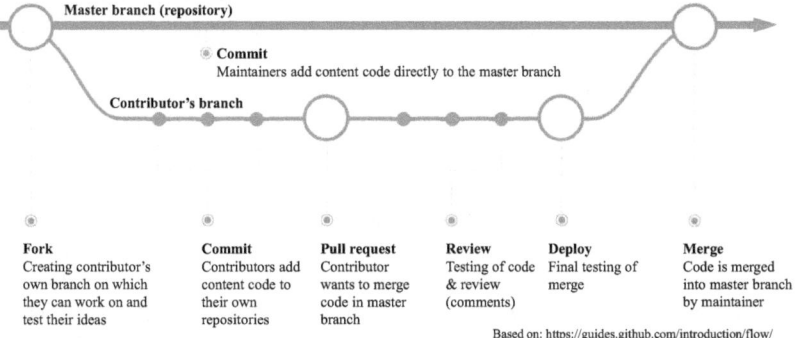

Figure 2.1 The GitHub workflow

---

[13] More complex OSS projects can be split up across several repositories, each containing modules of the overall OSS project. However, this is very rare.

The top of the figure depicts the development process in the master branch of the OSS project, which can be understood as the "official version" of the software hosted in the project owner's repository. Maintainers can directly make changes to the source code of the software. Such changes are called contributions or "commits" in GitHub. A commit can contain additional lines of source code, the deletion of lines of the source code, or modifications of already existing lines of code.

The bottom of the figure depicts the process of indirect contributions by contributors who do not have the rights of maintainers. Such contributors first need to create a *fork* of the main repository. Forks are copies of a repository and allow the contributor to freely experiment with changes in the source code without affecting the original repository.[14] Every GitHub user can make forks of any publicly hosted repository on GitHub without needing the permission of the project owner or its maintainers (Sims and Woodard, 2020). They can then make changes to the source code in their forked repository by means of commits. Once a contributor believes their changes should be included in the main project, they can make a *pull request* to the main repository. A pull request is a request that commits made to a forked repository are incorporated (or "pulled") back into the master branch (Sims and Woodard, 2020). The maintainers are then free to accept, ignore, or reject the proposed changes in a pull request (Tsay, Dabbish, and Herbsleb, 2014). Before making a final decision, they can review and discuss the changes with the contributor who made the pull request and other community members, ask the contributor for additional changes in the code, and test the code in the production environment ("deploy"). Once the maintainer has made its final decision to accept the proposed code changes of the contributor, the contributor's commits are *merged* into the master branch. The commits initially made to the forked repository can now also be found in the master branch, an important feature of GitHub, allowing to trace the commits' original creator.

Besides the "fork & pull" collaboration model, GitHub also allows further ways of social interaction on its platforms. For example, GitHub users are able to subscribe to updates by "watching" projects and "following" other GitHub users (Gousios and Spinellis, 2017). However, data on such activities is less granularly available for researchers (e.g., lacking time-stamped information), which is why they are not used in this thesis.

---

[14] The term 'fork' is traditionally negatively associated with OSS projects, as it implies that a developer has 'hijacked' a project and intends to take it in a direction that it will make it incompatible with the original OSS project (Sims and Woodard, 2020; Viseur, 2012). In GitHub, 'forking' has no negative connotation; it is simply a process enabling GitHub users to decouple their own work on a project from the original version (Sims and Woodard, 2020).

Lastly, I want to point out that this process of development in an OSS project is not an infinite, ongoing flow of contributions. Engagement in OSS projects is a highly dynamic process (Grand et al., 2004), and OSS projects may eventually lose traction in the community (i.e., community members stop making pull requests to the project). In fact, many OSS projects are short-lived (Liao et al., 2019). While research has widely covered why individuals and firms join OSS projects (see previous sections), there is limited research on factors influencing the health of OSS projects over time. So far, research has found that high activity levels in projects have a stabilizing effect (Oh and Jeon, 2007; Subramaniam, Sen, and Nelson, 2009), that the availability of other OSS projects can lead to decreasing contributions to the focal OSS project (Oh and Jeon, 2007), and that the availability of commercial, proprietary solutions can have a positive or negative impact on a project's health depending on the type of community members (Nagaraj and Piezunka, 2020). There have been mixed results on the influence of license types on the long-term health of OSS projects (Comino, Maneti, and Parisi, 2007; Subramaniam et al., 2009).

2.1.6 Summary

Section 2.1 introduced the core principles of OSS and the OSS development process based on informal collaboration among individuals and/or firms in OSS communities as an alternative to the proprietary software development process. It highlighted the history of OSS, the role of licenses for software to be defined as OSS, and the main two types of OSS licenses—copyleft and permissive licenses. It also highlighted the different benefits individuals and firms can obtain by participating in OSS development. While individuals participate in OSS development to gain intrinsic and extrinsic rewards, firms mostly do so for economic benefits, such as access to feedback and novel ideas, access to talent, fast adoption of their technology, and the potential to sell complementary services and products around their OSS. Particularly, firms also face potential disadvantages when open sourcing their technology: For example, other competitors can easily access and work with the firm's OSS at no cost, and firms may face difficulties in monetizing their OSS. Also, open-sourcing of a firm's technology in no way guarantees that a community will participate in the development of a firm's OSS project. As firms have no formal power over community members, the participants may freely choose which OSS projects they engage in and how. Thus, firms need to foster their relationship with the community. Lastly, the typical "fork & pull" collaboration model of OSS development was explained using GitHub as an example.

It was also noted that OSS development is a highly dynamic process, and many OSS projects are short-lived with factors, such as the overall project activity and the availability of alternative projects or proprietary alternatives influencing the chances of a project to "survive."

2.2   Basic Principles of Acquisitions and Theoretical Foundations

This section introduces the basic principles of acquisitions and related theoretical foundations. I will first explain the definition and characteristics of acquisitions (Section 2.2.1). I will then highlight the typical decision-making process related to an acquisition from an acquirer's perspective (2.2.2). I divide this process into three phases: The strategic planning phase, the pre-acquisition phase, and the post-acquisition phase. For each phase, I will highlight key elements and findings from research (2.2.3–2.2.5). Lastly, I will briefly summarize the key takeaways from this section (2.2.6).

As acquisitions are a multifaceted phenomenon (Arvanitis and Stucki, 2014), I cannot cover all aspects of acquisitions in detail in this chapter. Therefore, throughout the chapter, I will focus on the strategic aspects of managing the acquisition process and findings related to technology-focused acquisitions. I concentrate on technology-focused acquisitions since OSS development is a form of technological innovation. A major part of acquisition research has so far covered the role of proprietary innovation in acquisition decision-making and the impact of acquisitions on proprietary innovation. Furthermore, the relevance of technology-focused acquisitions has been increasing steadily in the past (de Man and Duysters, 2005; Kavusan, Ates, and Nadolska, 2020). I will, for example, barely cover the financial perspective of acquisitions as this stream of literature is not in the focus of this thesis (see, e.g., Aktas, de Bodt, and Cousin, 2007; Wilcox, Chang, and Grover, 2001; Haleblian and Finkelstein, 1999; Haspeslagh and Jemison, 1991).

2.2.1   Definition and Characteristics

Acquisitions are transactions on the market for corporate control (Manne, 1965; Sudarsanam, 2003) and describe the phenomenon of one firm—the acquirer—obtaining ownership of another firm's—the target's—stock, equity interests, or assets, and thereby obtaining control over the target. Acquisitions are an essential means to achieve inorganic growth and access resources and knowledge outside of the firm's boundaries to allow the firm to grow and evolve in today's dynamic markets. From the various options to source externally from the market, such as purchasing selected assets, licensing selected IP, joint ventures, or alliances, acquisitions are probably the biggest commitment out of all those options giving the acquirer full access to the employees, the IP, and possibly all existing products, services, and customers that the target might already have (Lambe and Spekman, 1997). Consequently, acquisitions are the most impactful form of external sourcing for an acquirer (e.g., Hlavka, 2019): Sourcing via an acquisition gives faster access to resources (Graebner, Eisenhardt, and Roundy, 2010) than a joint venture, which companies often form to develop new technologies, and unlike typical purchasing or licensing, acquisitions enable exclusive access to a target's resources and IP. In the case of acquisitions of targets active in OSS development, this would also include the copyright for OSS code the target possesses and the links to the communities the target engages with.

Acquisitions have a long history in strategic management, and activities have continuously increased in recent years (Kengelbach et al., 2017). Industries such as the software development sector, characterized by high complexity of the technology, fast-paced innovation, and highly specialized knowledge, regularly do not allow firms to internally develop all the capabilities and technologies required to stay competitive (Ranft and Lord, 2002). Acquisitions are thus central to keeping up with the competition. Consequently, acquisitions in digital technologies increased even more strongly than the overall acquisition activities in the last years (Kengelbach et al., 2017) and are of high relevance for both start-ups and incumbents (Thomson, Dettman, and Garay, 2018). Despite the high prevalence of acquisitions and some acquirers being highly successful in their acquisition efforts, success rates of acquisitions are generally poor, and it is regularly reported that, on average, somewhere between 40% and 90% of acquisitions fail to create value (Bagchi and Rao, 1992; Bower, 2001; Christensen, Alton, Rising, and Waldeck, 2011). Examining the conditions under which an acquisition generates value for the acquirer is therefore a key question for management research as an acquisition can be one of the most important decisions in a firm's history and can inadvertently affect its survival and success (Chondrakis, 2016; Dezi, Battisti, Ferraris, and Papa, 2018; Jensen, 1986; Moeller, Schlingemann, and Stulz, 2005).

Given their importance for a firm's survival and success, acquisition decisions are considered strategic decisions, which are most often in the responsibility of a company's top executives (Angwin, Paroutis, and Connel, 2015; Haspeslagh and Jemison, 1991; Jemison and Sitkin, 1986; Stein, 2017). Strategic decisions are characterized by their "novelty, complexity and open-endedness" (Mintzberg, Raisinghani, and Theoret, 1976) and "important, in terms of the actions taken, the resources committed, or the precedents set" (Mintzberg et al., 1976). The goal of strategic decision-making is to solve strategic problems—problems typically beset by risk, uncertainty and ambiguity (Pich, Loch, and De Meyer, 2002). The evaluation of decision alternatives, such as if an acquisition should take place, which target to acquire, or how to integrate the target, is therefore very challenging (Schwenk, 1984), and a large body of literature examines factors influencing uncertainty in acquisition decision-making (Capron and Shen, 2007; Chakrabarti and Mitchell, 2013; Fischer, Henkel, and Stern, 2020; Hernandez and Shaver, 2019; Hlavka, 2019; Ransbotham and Mitra, 2010; Rogan and Sorenson, 2014; Shen and Reuer, 2005; Zaheer, Hernandez, and Banerjee, 2010).

2.2.2   Brief Overview of the Typical Acquisition Decision-making Process

A brief overview of the typical acquisition decision-making process from an acquirer's perspective is given in the following. Understanding the acquisition process helps to comprehend the challenges an acquirer faces related to acquisitions, which are central to much of the management literature focusing on acquisitions. Furthermore, it has been shown that the acquisition process itself is a significant driver of the success of acquisitions (Jemison and Sitkin, 1986).

Extant literature has proposed anything from three (e.g., Meckl, 2004; Stein, 2017; Weber, Tarba, and Öberg, 2013) to seven phases (e.g., Angwin et al., 2015; Haspeslagh and Jemison, 1991) when describing the acquisition decision-making process. This thesis utilizes a simplified view consisting of three phases based on combining and simplifying work of Haspeslagh and Jemison (1991), Angwin et al. (2015), and Stein (2017). The process with the key steps within each of the three phases is depicted in Figure 2.2.

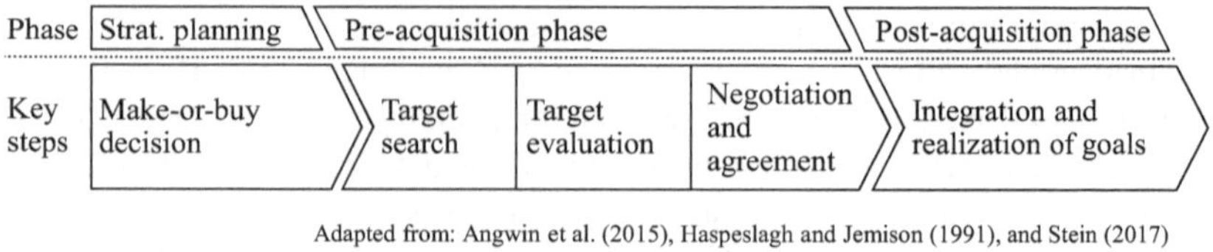

Phase	Strat. planning	Pre-acquisition phase			Post-acquisition phase
Key steps	Make-or-buy decision	Target search	Target evaluation	Negotiation and agreement	Integration and realization of goals

Adapted from: Angwin et al. (2015), Haspeslagh and Jemison (1991), and Stein (2017)

Figure 2.2 The typical acquisition decision-making process

These three phases are the strategic planning phase, the pre-acquisition phase, and the post-acquisition phase. For an acquisition to be successful, research showed that all phases need to be managed successfully (Barkema and Schijven, 2008; Bauer and Matzler, 2014; Stahl and Voigt, 2008). The strategic planning phase takes place at the acquirer internally and focuses on the strategic questions if and why an acquisition is necessary, and if the required resources should be bought via an acquisition or if they are better developed in-house (Chaudhuri and Tabrizi, 1999; Makri et al., 2010). This "make or buy" decision is relevant for allocating resources and is mainly driven by the internal gap in resources and capabilities and the competitive environment of the acquirer (Kurokawa, 1997; Makri, Hitt, and Lane, 2010; Stein, 2017). If an acquisition is desired, the strategic rationale for an acquisition, such as getting access to new products, services, technologies, or geographies, or mitigating a competitive threat, is a key outcome of the strategic planning phase (Galpin and Herndon, 2007; Stein, 2017; Weber et al., 2013). Furthermore, high-level target characteristics, such as technology, size, or position in the market are defined (Angwin et al., 2015; Stein, 2017; Weber et al., 2013).

The pre-acquisition phase includes all steps from searching for potential targets, via evaluating the targets and narrowing down the number of potential targets, to signing an acquisition agreement with one of them (see, e.g., Stein, 2017). Within these steps, strategic fit, cultural fit, and organizational fit, the potential targets' financial situation, and the potential targets' prior performance are evaluated, and due diligence is conducted (Angwin et al., 2015; Stein, 2017). Particular attention is paid to the potential targets' IP and to their research and development (R&D) teams in technology-focused acquisitions (e.g., Stein, 2017). The result of the pre-acquisition phase is an acquisition agreement with the owner of a target who is willing to sell (Galpin and Herndon, 2007; Weber et al., 2013). This also includes the acquisition price. After that, the legal transfer of ownership of the target can take place. The transfer of ownership is the pivotal moment in an acquisition and marks the separation of the pre- and post-acquisition phase (Gomes, Angwin, Weber, and Tarba, 2013).

After ownership has been transferred, the post-acquisition phase starts. In the post-acquisition phase, the acquirer seeks to realize value from the acquisition in line with the prior defined strategic rationale for the acquisition (Haspeslagh and Jemison, 1991). A key decision in this step is whether and how to structurally integrate a target into the acquirer. Integration gives the acquirer full access to and control over the acquired resources. However, particularly in technology-focused acquisitions, the target's technologies and capabilities are more difficult to transfer due to the tacitness and complexity of knowledge. Research has shown that particularly in such acquisitions, integration regularly destroys the same valuable capabilities that made the target attractive for the acquirer in the first place (Birkinshaw, Bresman, and Hakanson, 2000; Chaudhuri and Tabrizi, 1999; Graebner, 2004; Puranam, Singh, and Zollo, 2003; Puranam, Singh, and Chaudhuri, 2009; Ranft and Lord, 2002). If integration is desired, strategies need to be developed on how to integrate the firms organizationally and legally, on how to integrate the IT, and on how cultural change and adaption can be facilitated in the combined entity (Meckl, 2004).

After giving this brief overview of the acquisition decision-making process, selected key aspects of each phase relevant for this research will be presented.

2.2.3   Brief Overview of Strategic Acquisition Motives

A key step in the strategic planning phase of the acquisition decision-making process is the rationale or motive behind an acquisition. Often, the acquisition motive itself can have a significant impact on the chances of success of an acquisition (King, Dalton, Daily, and Covin, 2004; Rabier, 2017). According to general management theory, most acquisitions have the objective of enhancing firm competitiveness by accessing relevant resources and assets, redeploying them so they can be used more efficiently, and thus achieving better performance in the combined entity (Ahuja and Katila, 2001; Anand and Singh, 1997; Capron, Dussauge, and Mitchell, 1998; Dezi et al., 2018; Haspeslagh and Jemison, 1991). Beyond this general goal of gaining control and redeploying assets as a lever for enhancing competitiveness, research recognizes several strategic motives behind acquisitions. While there is no broadly accepted list of acquisition motives, an overview of key acquisition motives identified across the management literature will be provided in the following. In the context of this thesis, a general understanding of acquisition motives is a foundation for the analysis of why firms active in OSS get acquired. Furthermore, acquisition motives are a key driver of the outcomes of acquisitions.

Access to resources and technologies. A main driver of many acquisitions is the desire to get access to a target's resources, including technologies, capabilities, sales networks, and manufacturing capacities possessed by the target (Ahuja and Katila, 2001, Chaudhuri and Tabrizi, 1999; Ranft and Lord, 2002; Worek et al., 2018). Compared to other means to get access to resources, acquisitions offer exclusive access to a target's resources and allow to get access faster than building those resources internally, which is particularly relevant in fast-paced industries like the technology or software sector (Puranam and Srikanth, 2007;

Warner, 2003). In some cases, acquisitions may be the only way to get access to specific resources, for example, if the target firm does not want to sell or provide access to those otherwise.

A particularly relevant form of such resource-focused acquisitions are technology-focused acquisitions, where the acquirer seeks access to a target's technologies and IP to bolster its innovation capabilities (Giuri et al., 2006). This motive appears to have increased in relative importance since the 1990th and acquisitions of small, technology-intensive targets are a major source of renewal and growth in industries with fast-paced technological change (Bower, 2001; Karim and Mitchell, 2000; Kengelbach et al., 2017). Such technology-focused acquisitions are attractive means for enhancement of existing capabilities and technological diversification and allow acquirers to quickly jump onto new technology trends or bridge gaps in their technological capabilities (Capron and Mitchell, 2009; Hussinger, 2010). They provide technological input to the acquirer enabling it to avoid the uncertainty coming with internal technology development and to instead leverage the target's existing technologies (Ahuja and Katila, 2001; Puranam and Srikanth, 2007). However, acquisitions can only partly replace internal technology development, and researchers found that acquiring technologies and internal technology development are best understood as complements (Cassiman and Veugelers, 2006; Granstrand, Bohlin, Oskarsson, and Sjöberg, 1992). A further distinction within technology-focused acquisitions was made by Stein (2017), who differentiated technology-focused acquisitions in acquisitions aiming to broaden the functionality of an acquirer's technology or product and acquisitions aiming to enhance the performance of an acquirer's technology or product.

Acqui-hire. Acquisitions aiming to acquire a target's (highly skilled) human resources are a special form of resource-focused acquisitions often recognized as a separate acquisition motive (Coyle and Polsky, 2013; Kim, 2019; Ouimet and Zarutskie, 2012). In such acquisitions, acquirers desire to obtain the target's in-depth experience and skills of specific technical and sometimes managerial employees (Ranft and Lord, 2002). In these acquisitions, acquirers are less likely to be interested in a target's physical resources like infrastructure, plants, or equipment, or IP like licenses; instead, the acquirers desire to obtain intangible knowledge difficult to build internally. This practice is prevalent for acquiring start-ups whose most valuable—often the only—assets are their employees (Chatterji and Patro, 2014; Kim, 2019).

Efficiency gains. Acquirers to realize efficiency gains has been a historically important acquisition motive. Acquirers engage in such acquisitions to achieve synergies they foresee between the acquirer and the target (Trautwein, 1990). The desire is that the combined entity can benefit from economies of scale and scope (Cassiman, Colombo, Garrone, and Veugelers, 2005) and that duplicated efforts can be avoided (Veugelers, 2006). While economies of scale reduce the average cost of production through the increased size of the combined entity, economies of scope are complementarities that make it cheaper to produce products jointly rather than to produce each of them individually (see, e.g., Chakrabarti, Hauschildt, and Sueverkruep, 1994; Seth, 1990; Sharma and Ho, 2002).

Market expansion. Acquisitions with the desire to expand the market refer to acquisitions aiming to access new customers in markets new to the firm (Cloodt, Hagedoorn, and Van Kranenburg, 2006). Acquirers involved in such acquisitions either aim to get access to customers in new product markets or in new geographic markets (Stein, 2017; Worek et al., 2018). In the first case, acquirers aim to diversify their product portfolio by selling products with little or no connection to their existing business (Bower, 2001; Stein, 2017). In the latter case, acquirers instead seek to extend their core business to another geography (Bower, 2001; Stein, 2017). In both cases, firms reduce their risk due to diversification.

Portfolio expansion. In contrast to the market expansion motive, the portfolio expansion motive aims to extend the product portfolio with complementary products or services (Cloodt et al., 2006; Worek et al., 2018). Acquiring targets with complementary products or services allows the acquirer to generate additional revenue streams by cross-selling the products to existing customers. Market and portfolio expansion motives together can be referred to as general expansion motives, which cater to a firm's desire to grow and to commercialize (new) products and services (Worek et al., 2018).

Pre-emption of competition. Acquirers might also acquire targets to shape their (future) competitive environment and reduce potential competition. Several variations of this motive have been identified by prior literature. One variation is the "buy before someone"-acquisition motive described by Fridolfsson and Stennek (2005), where the acquirer seeks to acquire a target to avoid one of the acquirer's competitors doing so. Another one is the "killer acquisition"-acquisition motive described by Cunningham et al. (2020), where a start-up is acquired, and its product or technology stopped from being developed further in order to avoid future competition from the start-up. Lastly, the increased market power of the combined entity after the acquisition can also increase barriers to enter the market and as a result mitigate potential competitors entering this market (Comanor, 1967). In this context, Grimpe and Hussinger (2008) have shown that firms possessing technologies with the potential to deter market entry are of high value to acquirers.

Financial motives. Financial considerations might also drive acquisitions. Acquirers might acquire firms to realize financial synergies, such as tax savings, lower cost of capital, and diversification of cash flow streams (Rabier, 2017). Acquisitions may also be opportunity-driven; for example, when an acquirer buys a target it considers to be undervalued (Angwin, 2007; Wernerfelt, 1984).

Other motives. The list of acquisition motives above is by far not exhaustive, and other authors have created similar lists. Other motives for acquisitions mentioned in the literature are cultural fit (Worek et al., 2018), political reasons (Chakrabarti et al., 1994), or hubris of managers (Roll, 1986). For example, Worek et al. (2018) provide a list of 46 different acquisition motives.

To conclude this overview of acquisition motives, two important aspects of acquisition motives should be highlighted: First, in most acquisitions multiple motives are involved (Graebner et al., 2010; Worek et al., 2018). Second, not only acquirers need a motive for an acquisition to take place, but also the owners of a target need a motive to sell their firm. An acquisition can only occur if an acquirer desires to acquire a target whose owner is willing to sell (for an overview of such motives, see Graebner and Eisenhardt, 2004).

2.2.4 Key Aspects of and Relevant Research on the Pre-acquisition Phase

Key tasks in the pre-acquisition phase are to find potential targets, select the most suitable one for an acquisition, and then settle on the terms of the deal with the target's owner. Choosing the right target is crucial for acquisition success (Haspeslagh and Jemison, 1991). However, making acquisition decisions is not easy and poses many challenges to acquiring managers as the complexity of strategic, organizational, and cultural challenges is typically higher than the complexity of, for example, internal investment decisions (Haspeslagh and Jemison, 1991; Pablo, 1996). In the following, key challenges acquirers face when searching, selecting, and evaluating potential targets in the pre-acquisition phase, and key findings from research relevant for these processes will be highlighted.

Key challenges acquirers face in the pre-acquisition phase
In the target search process, acquirers need to find potential targets. Research has shown that this search process is limited by the time acquiring managers spend on searching, as acquiring managers have limited resources to do so (Chakrabarti and Mitchell, 2013). As a result, the target search process typically only considers a fraction of the population of potential targets (Chakrabarti and Mitchell, 2013).

Once a list of potential targets exists, they need to be evaluated, and the list needs to be boiled down to one or a few selected candidates to conduct detailed due diligence and enter formal negotiations with the owner of the target(s). For both, the search and particularly the evaluation and selection of targets, acquirers need extensive information about the potential targets (Chakrabarti and Mitchell, 2013). This includes financial information such as a target's profitability, debt, the value of the target's assets, and the financial resources the acquirer will be required to invest in the target in order to upgrade, redeploy, or utilize the targets' resources after the acquisition (Chakrabarti and Mitchell, 2013; Salter and Weinhold, 1981; Weston, Mitchell, and Mulherin, 2004). In order to assess the cultural fit of the firms, acquirers also need information on the culture of the targets (Chakrabarti and Mitchell, 2013). Furthermore, they need information on the targets' organization, including corporate responsibilities, employment policies, and benefit plans (Yunker, 1983). Acquirers have to evaluate how a targets' capabilities fit to their own organization's (Haspeslagh and Jemison, 1991) as well as which effort will be required to successfully combine the acquirer's and the target's resources (Capron et al., 1998).

It becomes clear that getting access to and collecting such information poses a significant challenge for acquirers. Furthermore, many of the dimensions on which acquirers would like to evaluate potential targets remain difficult to assess even after thorough due diligence as acquirers face an information asymmetry (Chondrakis, 2016; Hansen, 1987; Rogan and Sorenson, 2014). Founders, owners, and managers of a potential target understand the firm's strengths and weaknesses better than the acquirer (Rogan and Sorenson, 2014). They might also understand the market environment they operate in better than an acquirer not yet active in this product- or regional market. However, those founders, owners, and managers regularly have their own goals, and might even wish to promote a poor acquisition before evidence of the flaws of the target becomes apparent to the acquirer. As a result, acquirers cannot only rely on information they receive from the targets (Rogan and Sorenson, 2014).

Overall, the pre-acquisition phase of acquisitions is therefore characterized by a lot of uncertainty the acquirer faces about its potential targets (Ransbotham and Mitra, 2010; Rogan and Sorenson, 2014; Zaheer et al., 2010). Research distinguishes four types of uncertainty in acquisition decision-making: Market risk, technology risk, integration risk, and financial risk (Chaudhuri, Marco, and Tabrizi, 2005; Stein, 2017). Market risk refers to the question of whether or not a target's customers will adopt the technology, if it fits customer needs, if customers are willing to buy the product at a sufficient price, and if there will be any customers at all (MacCormack and Verganti, 2003; MacMillan and McGrath, 2002). Furthermore, it includes whether the revenues can be scaled in the future and how competitors react to the target and a potential acquisition (Stein, 2017).

Technology risk refers to whether or not the target's technological inventions will perform as desired (Chaudhuri et al., 2005). This includes risks such as that a technology development project is not coming to fruition due to obstacles that cannot be overcome, and thus an invention will vanish without being applied (Chaudhuri et al., 2005; MacCormack and Verganti, 2003; MacMillan and McGrath, 2002). It also encompasses compatibility, cybersecurity, scalability, standard-setting, and time-to-market risk (Stein, 2017). Integration risk refers to whether or not the target and the acquirer can be integrated after the acquisition in the way the acquirer intends to. This is particularly relevant, as integration regularly destroys the same valuable capabilities that made the target attractive for the acquirer in the first place (Birkinshaw et al., 2000; Chaudhuri and Tabrizi, 1999; Graebner, 2004; Ranft and Lord, 2002; Puranam et al., 2003; Puranam et al., 2009). Particularly the cultural fit of both firms and the risk of acquired employees leaving the combined entity after an acquisition are relevant for the integration risk (Kim, 2019). Lastly, financial risk refers to the risk for the acquirer to lose the invested capital (Stein, 2017). It increases with deal value and is the product of market risk, technology risk, and integration risk and is therefore not as fundamental as the first three (Stein, 2017).

Research has found that younger potential targets are associated with more uncertainty than older ones, and that information asymmetry between target and acquirer are particularly present for younger targets (Hussinger, 2010; Shen and Reuer, 2005): for younger targets it is less clear whether its product or technology will eventually succeed and if there is actually a market for the product or technology.

Research findings on key aspects of pre-acquisition phase decision-making
The aforementioned uncertainty and information asymmetries acquirers face in their target search, selection, and evaluation processes are key factors influencing the decision-making process and have therefore been in the focus of management research in the past. Research found that acquirers value information that helps to decrease uncertainty about a potential target's quality. But other factors have also been found to influence target search, selection, and evaluation. In the following, key findings from research on three key decisions that an acquirer needs to make within target search, selection, and evaluation will be highlighted, for which research has identified several factors that influence those areas. The three key decisions are the target selection from a set of potential alternatives, the timing of acquisitions, and deal values. Key insights across the three decisions will be summarized at the end of this section.

Target selection. Target selection research focuses on the question of which target gets acquired among a set of potential alternatives. Several researchers have found that acquirers value information that helps to reduce uncertainty about a potential target's quality and increases the likelihood that targets signaling higher quality are selected. Means to do so are, for example, patents (Ransbotham and Mitra, 2010), which signal technical quality or prior collaboration of the target and the acquirer (Shen and Reuer, 2005). Next to collaboration, other direct or indirect ties between them, for example, common clients or a supplier-customer relationship, can also increase the probability of the potential target to get acquired (Hernandez and Shaver, 2019; Rogan and Sorenson, 2014; Zaheer et al., 2010). Acquisition likelihood also increases when technological uncertainty declines due to regulatory approval of a new product (Fischer et al., 2020) or when changing regulation requires firms to publish more information about themselves (Chondrakis, Serrano, and Ziedonis, 2020). Research also found that—given the presence of public and private alternatives—acquirers tend to choose public targets over private targets, particularly if the industry is new for the acquirer or the targets have significant intangible assets, something commonly observed in the software industry (Capron and Shen, 2007). Additional factors having the potential to reduce uncertainty about targets are signals from Initial Public Offering (IPO) processes, business analysis, connections to investment banks, and press coverage (Chakrabarti and Mitchell, 2013; Reuer and Ragozzino, 2008)

Collecting information on targets becomes more difficult with an increasing geographic distance between the target and the acquirer, resulting in a lower probability of an acquisition to take place (Chakrabarti and Mitchell, 2013). In the context of geographic distance, acquirers seem to prefer inferior targets in existing geographic markets but are more likely to choose superior targets in new markets (Kaul and Wu, 2015). Acquisition likelihood is also positively related to technological similarity and complementarity (Kavusan et al., 2020), and acquirers prefer similar targets when it comes to R&D pipelines, while they prefer complementary targets when it comes to product portfolios (Yu, Umashankar, and Rao, 2016). If potential targets publish more information about themselves, the average technological distance between target and acquirer increases (Chondrakis et al., 2020). Additional factors that have been shown to influence acquisition likelihood positively are a favorable position of the potential target in a network of firms (Hernandez and Shaver, 2019) and similarity in national culture between the acquirer and the potential target (Rao, Yu, and Umashankar, 2016).

Acquisition timing. Acquisition timing research focuses on the question of the age of a target at the time of the acquisition—i.e., whether a target is acquired earlier or later in its life cycle (e.g., Hlavka, 2019; Ransbotham and Mitra, 2010;

Stein, 2017). Similar to target selection research, research has found that uncertainty is a key factor influencing acquisition timing. Acquirers face a trade-off between uncertainty and higher acquisition prices when deciding if they want to acquire a target earlier or later (Sorenson and Stuart, 2008). The earlier in its lifecycle a target is bought, the less expensive it usually is. However, the younger the target, the less clear it is whether the target's product or technology will eventually succeed and if there is a market for the product or technology. As a result, younger targets are associated with more uncertainty than older ones (Shen and Reuer, 2005). When deciding when to acquire a target, acquirers might choose to delay acquisitions until the associated risks have reached a certain acceptable threshold (Alvarez and Stenbacka, 2006).[15] Particularly for technology-focused acquisitions, waiting is a legitimate strategy for obtaining additional information to reduce uncertainty (Stein, 2017). In turn, factors reducing uncertainty in acquisition decision-making can lead to earlier acquisitions. In this context, Hlavka (2019) found that acquirers' capabilities to evaluate targets, such as experience from previous acquisitions, can reduce uncertainty for the acquirer and lead to younger targets. Research has also shown that in the market for novel technologies, followers get acquired earlier relative to their founding date than first movers, as they can benefit from the progress the first mover has made in reducing the uncertainty surrounding a novel technology (Fischer et al., 2020). Lastly, Alvarez and Stenbacka (2006) found that acquisitions take place earlier with decreasing uncertainty on when synergies can be realized after an acquisition.

But not only factors reducing uncertainty influence acquisition timing. Hlavka (2019) found that acquirers shift towards earlier acquisitions despite increasing uncertainty around a target during technology hypes. Furthermore, Stein (2017) found that acquisition targets are younger and smaller in technology-focused acquisitions focusing on performance improvement of the acquirer's technology portfolio compared to technology-focused acquisitions focusing on adding additional functionalities to the acquirer's technology portfolio. Lastly, acquirers also tend to buy targets possessing certain technologies the acquirer currently lacks earlier if they expect this technology to become part of a standard (Warner et al., 2006).

---

[15] One should note that the acquirer cannot alone decide on acquisition timing—the prospective seller most often has a say in this decision as well. For example, if the potential seller is more optimistic about its firm's future value, it will prefer to delay the acquisition until uncertainty about its value is reduced rather than accepting a lower price (Allain, Henry, and Kyle, 2016).

Deal value. The last aspect that should be covered are deal values—i.e., the prices acquirers pay for targets. The deal value is a key outcome of the pre-acquisition decision-making. It can be considered a proxy for the acquirer's "willingness to pay" (Brandenburger and Stuart, 1996). It reflects the acquirer's expectations about the acquirer's possibilities to create and capture value in the future (Barney, 1988). However, particularly in knowledge-intensive industries acquirers face a high risk of overpaying caused by uncertainty and asymmetric information between target and acquirer (Coff, 1999). A substantial amount of research from a financial perspective has indicated that "paying too much" is a major cause of the failure of acquisitions (see Gomes et al., 2013, for an overview of respective finance literature). A too high deal value may have a negative impact on the post-acquisition performance, as it becomes difficult to achieve an adequate return on the investment (Datta, 1991; Goold, Campbell, and Alexander, 1994; Haunschild, 1994; Hayward and Hambrick, 1997).

A way to decrease the risk of overpaying is to seek better information about the target (Coff, 1999). Waiting is another option; while prices regularly increase with growing target age, uncertainty around its quality decreases, making it easier to assess its true value. For example, Reuer and Ragozzino (2008) highlight the case of eBay acquiring PayPal only after PayPal went public and the equity market confirmed the value PayPal claimed it was worth. In this context, it has been shown that private targets, for which less information is available, regularly sell with a substantial price discount reflecting the information discount that acquirers experience while evaluating such targets (Chakrabarti and Mitchell, 2013; Koeplin, Sarin, and Shapiro, 2000).

Summary. In summary, this section highlighted key challenges acquirers face in decision-making on the pre-acquisition phase and factors having an impact on the decision-making process. Particularly the role of uncertainty about a target's qualities and information asymmetry between target and acquirer are omnipresent in the pre-acquisition phase. High uncertainty may lead to deferring acquisitions, potentially overpaying the seller, or not selecting a firm as a target altogether. Research has shown several factors that help acquirers reduce uncertainty in acquisition decision-making. For example, collaboration with the target, the publicly known patent stock of targets, or regulatory approval of a target's innovation can help acquirers reduce uncertainty and lead to better decisions in the pre-acquisition phase.

2.2.5   Key Aspects of and Relevant Research on the Post-acquisition Phase

The financial and legal aspects of an acquisition are often well handled, but managers regularly fail to successfully organize and manage the post-acquisition phase (Gomes et al., 2013). However, this phase is the most important one in an acquisition, as it is the phase where the value is created (Haspeslagh and Jemison, 1991). Research estimates that somewhere between 40% and 90% of acquisitions fail to create value (Bower, 2001; Christensen et al., 2011; Bagchi and Rao, 1992). This is also true for technology-focused acquisitions, which regularly do not meet the expectations (Bannert and Tschirky, 2004; Chaudhuri and Tabrizi, 1999; Puranam and Srikanth, 2007; Steensma and Corley, 2000; Stein, 2017) and failure rates of up to 80% have been observed (Puranam and Srikanth, 2007). Research has focussed on two key aspects of the post-acquisition phase, namely the measurement of the impact of acquisitions on post-acquisition performance and the drivers of success in the post-acquisition phase. Regarding the latter, one can further differentiate prior research on the role of the strategic fit of the acquirer and the target based on pre-acquisition variables for post-acquisition performance and the role of the management of the post-acquisition phase for post-acquisition performance. In the following, key findings from research in these areas will be highlighted.

Measurement of the impact of acquisitions
There are a number of different ways to measure the impact of acquisitions. Traditional measures for the impact of acquisition are accounting-based, like revenue growth, or based on stock returns, for example, in event studies (Haleblian and Finkelstein, 1999; Larsson and Finkelstein, 1999; Meglio, 2009). For example, the latter measure how the market reacts to acquisitions, i.e., how acquisition announcements of public firms are perceived by investors and thus produce positive or negative abnormal returns. However, management literature also recognizes a number of other approaches, such as measuring changes in the organizational structure of the target (Karim, 2006), changes in the behavior and culture (Birkinshaw et al., 2000; Nahavandi and Malekzadeh, 1988), employee turnover after an acquisition (Hambrick and Cannella 1993; Ernst and Vitt, 2000; Krug and Nigh, 2001; Kim, 2019), or subjective perception of the acquisition success based on surveys or interviews (Meglio, 2009; Nielsen and Gudergan, 2012). More recently, research also included reactions of other stakeholders such as customers or competitors when measuring the impact of acquisitions (Kato and Schoenberg, 2014;

Rogan and Greve, 2015; Valentini, 2016). Measures frequently used when studying technology-focused acquisitions focus on innovation success, a measure often
based on patent quantity or patent quality[16] (Ahuja and Katila, 2001; Valentini,
2012) or the number of new product introductions (Puranam, Singh, and Zollo,
2006).

Particularly in the area of technology-focused acquisitions, many authors
report acquisitions having a negative impact on innovation (e.g., Ernst and Vitt,
2000; Hitt, Hoskisson, and Ireland, 1990; Hitt, Hoskisson, Ireland, and Harrison,
1991; Kapoor and Lim, 2007; Ranft and Lord, 2002). Similarly, key acquired rese-
arches often leave the target after an acquisition (Cunningham et al., 2020; Ernst
and Vitt, 2000; Kim, 2019). Potential causes for the negative impact are missing
managerial focus (Hitt et al., 1990), reduced investment in innovation (Hitt et al.,
1991), disruption of routines (Ranft and Lord, 2002), misalignment of incentives
(Kapoor and Lim, 2007), and uncertainty (Ernst and Vitt, 2000). Besides many
authors reporting a negative impact of acquisitions on post-acquisition innovation
outcomes, some authors report mixed results (e.g., Ahuja and Katila, 2001; Cassiman et al., 2005; Prabhu, Chandy, and Ellis, 2005; Valentini, 2012). For example,
Valentini (2012) found that acquisitions have a positive effect on overall patenting output, but a negative effect on patent impact, originality, and generality.
Lastly, a number of authors stress that the strategic rationale and the acquirers'
intentions need to be considered when evaluating acquisition success—in particular whether the acquirer intends to redeploy or divest resources following the
acquisition (Bannert and Tschirky, 2004; Capron, Dussauge, and Mitchell, 1998;
Capron, Mitchell, and Swaminathan, 2001; Karim and Capron, 2016; Ranft and
Lord, 2002). For example, in an extreme case, the acquirer might want to "kill" a
potential competitor's innovation (see Cunningham et al., 2020). In this case, the
negative impact on post-acquisition performance might well be exactly what the
acquirer desired.

However, such "killer acquisitions" are rare, and in most technology-focused
acquisitions the acquirer indeed wants to boost its innovativeness. Hence, a large
body of research focuses on understanding drivers of post-acquisition success. To

---

[16] Patents are probably the most used measure for post-acquisition innovation output in
technology-focused acquisitions. Yet, they have strengths and weaknesses (Bauer, Matzler,
and Wolf, 2016). They are directly linked to inventiveness of a firm (Ahuja and Katila, 2001)
and provide a measure for technical knowledge (Prabhu, Chandy, and Ellis, 2005). Yet, patents
are only capturing codified knowledge, some inventions are not patentable, and value of patents
is difficult to assess in terms of a cash price (Bauer et al., 2016). Furthermore, the measure
most often does not account for a firm's propensity to patent, i.e., the measure does not capture
firms not filing patent applications for strategic reasons.

understand the mechanisms behind contingencies in the outcomes of acquisitions, research focuses on the strategic fit of the acquirer and the target based on pre-acquisition variables and how the post-acquisition phase is managed. Key findings from both perspectives are covered in the following.

Strategic fit of acquirer and target

The strategic fit of target and acquirer based on pre-acquisition variables is considered a key factor for acquisition success in strategic management (Bauer and Matzler, 2014; Homburg and Bucerius, 2006; King et al., 2004; Seth, 1990). Particularly important is the "fit" of the acquirer's and the target's resources. Here, one needs to further differentiate strategic similarity and strategic complementarity. Researchers regularly argue that similarity—for example, market, resource, technology, and/or supply chain similarity—is an indicator of the synergy potential in an acquisition (Bauer and Matzler, 2014). It is assumed that similarity has a positive effect on post-acquisition success via increased market power, economies of scale and scope, and the minimization of redundancies (Bauer and Matzler, 2014). While higher similarity seems to be associated with better outcomes (Capron, Mitchell, and Swaminathan, 2001; Prabhu et al., 2005; Swaminathan, Murshed, and Hulland, 2008; Tanriverdi and Venkatraman, 2005), there are mixed results (Bauer and Matzler, 2014). Several researchers argue that complementarity of the acquirer's and the target's resources are more relevant for post-acquisition success than the similarity of their resources (Ahuja and Katila, 2001; Cassiman et al., 2005). Complementary resources of acquirer and target offer the possibility of valuable resource combinations between target and acquirer (Larsson and Finkelstein, 1999). In this case, there are different value creation mechanisms in place. Whereas similarity is interpreted as an indicator for efficiency-based synergies, complementarity provides companies with both efficiency increases based on synergies and value created from differences of the acquirer's and the target's resources that are mutually supportive and enhancing (Bauer and Matzler, 2014). For technology-focused acquisitions, this argument is supported by the findings of Ahuja and Katila (2001), who found that acquisitions of targets with a complementary technology base resulted in higher levels of post-acquisition innovation compared to acquisitions of targets with a very similar or unrelated technology base. Similarly, Makri et al. (2010) show that the quality of innovations produced after an acquisition is positively influenced by technological complementarity but not by technological similarity. Furthermore, technological similarities facilitate incremental renewal, whereas technological complementarities facilitate strategic renewal. Furthermore, Colombo and Garrone (2006) show that technological complementarity leads to increased investments in R&D, while technological similarity leads to decreasing investments in R&D.

Besides the similarity and complementarity of resources, also other factors need to be considered. Particularly organizational fit and cultural fit are important (Ernst and Vitt, 2000; Bauer and Matzler, 2014). For example, similar cultures—literature further differentiates national, regional, industrial, and professional culture (Gomes et al., 2013)—may help to facilitate a smooth integration and to retain the target's talent after an acquisition (Ernst and Vitt, 2000; Bauer and Matzler, 2014), while a significant size difference of acquirer and target can have a negative impact on post-acquisition innovation (Cloodt et al., 2006). Furthermore, the prior acquisition experience of the acquirer can positively influence post-acquisition performance (Barkema and Schijven, 2008).

Overall, studies found that if both, fit of the firm's resource and organizational fit, are known before the acquisition and are taken into account in the post-acquisition phase, acquisition performance is superior to those acquisitions where pre-acquisition characteristics are not considered for post-acquisition decision-making (Gomes et al., 2013).

Management of the post-acquisition phase

Not only the fit of target and acquirer has an influence on the outcomes of acquisitions, but also the management of the post-acquisition phase by the acquirer is crucial. Key factors regarding the management of the post-acquisition phase that are considered critical for post-acquisition success are the level of structural integration, the speed of integration, the post-acquisition leadership, and the communication of the process (Gomes et al., 2013).

Structural integration of the target into the acquirer's organization is regularly required in acquisitions to exploit potential synergies between target and acquirer (Gomes et al., 2013). The term "structural integration" describes the combination of the target's and the acquirer's organization into the same organization after the acquisition (Haspeslagh and Jemison, 1991; Paruchuri et al., 2006; Puranam et al., 2006; Puranam and Srikanth 2007; Puranam et al., 2009). Without any integration it is not possible to redeploy resources and reduce redundant resources (Bauer and Matzler, 2014), and a lack of integration can be a reason for acquisitions to fail (Schweiger and Weber, 1989). Yet, the degree of integration is a mixed blessing (Bauer and Matzler, 2014). On the one hand, tight integration can be necessary for realizing synergies between target and acquirer and therefore necessary for post-acquisition success. On the other hand, besides the direct cost of increased coordination, restructuring, and redeployment of resource (Bauer and Matzler, 2014; Shrivastava, 1986), integration can cause organizational disruption, and the loss of autonomy following an acquisition can cause demotivation of the acquired employees (Aghasi, Colombo, and Rossi-Lamastra, 2017; Datta, 1991; Datta and Grant, 1990; Larsson and Finkelstein, 1999; Ranft and Lord, 2002). Particularly in the presence of cultural differences, much integration can be detrimental to the outcomes of acquisitions as the potential for a clash of target's and the acquirer's culture is higher (Gomes et al., 2013; Weber and Schweiger, 1992).

A high degree of integration is required to realize the anticipated benefits of technology-focused acquisitions, as such acquisitions require to obtain and transfer tacit and socially complex knowledge-based resources (Puranam et al., 2003; Ranft and Lord, 2002). However, prior research has often found a negative impact of integration in acquisitions on post-acquisition innovation, highlighting the difficulty to successfully transfer such knowledge (e.g., Paruchuri et al., 2006; Puranam et al., 2006). Reasons identified for the negative impact of integration on post-acquisition innovation are high employee turnover after acquisitions, disruption of organization routines, and dissolution of embedded ties in the target resulting in lower productivity of the employees (Ernst and Vitt, 2000; Gomes et al., 2013; Paruchuri et al., 2006; Puranam et al., 2006). This highlights a key dilemma in when it comes to integrating targets—the extent of structural integration needed to achieve the desired synergies may result in the destruction of the target's knowledge-based resources (Gomes et al., 2013).

Another aspect of integration relevant to the management of the post-acquisition phase is the speed of integration. Higher speed reduces the time to reach the desired degree of integration after the acquisition deal is closed (Homburg and Bucerius, 2006). It is argued that a high speed of integration can lead to faster realization of synergies and allows to minimize time spent in a suboptimal condition (Angwin, 2004; Bauer and Matzler, 2014). Furthermore, uncertainty among employees can be reduced through fast integration (Angwin, 2004; Homburg and Bucerius, 2005). Consequently, a higher speed of integration can result in earlier returns on the acquirer's investment (Angwin, 2004). However, a lower speed of integration may also have benefits. Researchers highlighted that slow integration helps build trust among company employees and can help to reduce conflicts in the integration phase (Olie, 1994; Ranft and Lord, 2002). Hence, as Gomes et al. (2013) concludes, "there is no 'right' speed at which to perform the integration process." Managers need to balance the trade-off of the potential advantages and potential downsides when choosing the speed of integration.

Besides integration, other factors have been found to be crucial for managing the post-acquisition phase. One factor of relevance is the leadership style in the post-acquisition phase. To ensure a successful completion of the acquisition, the top management must take decisive action and establish a clear direction for the target (see Gomes et al., 2013, for a more detailed overview). Another factor, and the last one I want to cover in this overview, is the communication in the post-acquisition phase. To minimize the effects of insecurity and uncertainty among a company's stakeholders in the post-acquisition phase, it is crucial to deal with the stakeholder's anxieties utilizing various communication channels (Gomes et al., 2013; Schweiger and DeNisi, 1991). A particular focus needs to be put on the human resources of the involved firms (Ranft, 2006).

Summary. This section described different approaches to measure the impact of acquisitions on firm performance (e.g., stock market return-based measures) or innovation (e.g., patent-based measures) and highlighted that research often finds that acquisitions frequently fail to achieve performance improvements. Contingencies can be explained with the strategic fit of the acquirer and the target and how the post-acquisition phase is managed. For both areas, researchers have identified several factors of relevance. Particularly resource complementarity and cultural fit are important for the strategic fit of firms. For the management of the post-acquisition phase, the level of structural integration and the management of the integration process are key.

2.2.6 Summary

This brief overview introduced the basic principles of acquisition and theoretical foundations. An acquisitions is a transaction on the market for corporate control, where one firm—the acquirer—purchases control over another one—the target. The acquisition decision-making process consists of three phases: The strategic planning phase, the pre-acquisition phase, and the post-acquisition phase. In the strategic planning phase, the acquirer needs to decide whether an acquisition is the right measure to achieve the desired goals. Section 2.2.3 highlighted the most relevant rationales for acquisitions. In the pre-acquisition phase, the acquirer needs to search potential targets, evaluate them, and select one of them. Once the deal is closed, the post-acquisition phase begins. The post-acquisition phase is the phase where value is created by implementing the desired strategy based on the goals. Integration and reconfiguration of resources are often necessary to realize potential gains. However, many acquirers fail to realize the value they have hoped to gain in acquisitions. Reasons can be a suboptimal fit of target and acquirer or suboptimal management of the post-acquisition phase.

2.3   Intermediate Conclusion: Two Perspectives Bringing OSS and Acquisition Research Together

This chapter presented the basic principles and theoretical foundations of both, OSS development and acquisitions. The first highlighted the differences of OSS as an open, collaborative form of innovation characterized by informal collaboration compared to proprietary software development. It also showed that firms increasingly participate in OSS development and shed light on their motivation to do so. The latter highlighted the different phases of the acquisition process and key decisions in each phase. The goal of this thesis is to bring both research fields together.

Very simplified, the acquisition process can be grouped in a phase before the actual transaction takes place and a phase after the transaction. In both phases, OSS might play a role, and both roles should be covered in this thesis. This section will briefly show how the two research objectives formulated in Chapter 1—each focusing on one phase of the acquisition process—bring both literature streams together and elaborate the structure of the following chapters, where the research questions are tackled.

2.3.1   OSS Development as an Input to the Strategic Decision-making Process before an Acquisition

The first perspective on the connection between OSS development and acquisitions focuses on the strategic decision-making process leading to acquisitions. In the decision-making process leading to acquisitions, I conceptualize information related to the targets' and their acquirers' OSS activities as an input to the decision-making process. In this phase of the acquisition process, acquirers will need to think about if and why they acquire a target active in OSS—after all, the OSS itself is most often available for free on the web. Furthermore, the fact that OSS development takes place in the open potentially gives new opportunities for acquirers to evaluate a target based on its OSS activities—something not possible the same way for firms focusing on proprietary software development or other forms of proprietary innovation, as those innovation processes and their output is typically kept secret or at least protected. Hence, activities of potential targets in OSS development might also influence target search, evaluation, and selection processes. Prior literature has neither covered OSS-specific acquisition motives, nor the role of OSS, or openness in general, for target search, evaluation and selection processes. As formulated in Chapter 1, the first research objective connecting OSS development and acquisitions therefore is:

- **Research objective 1**: Create an understanding of the role of targets' OSS activities for their acquisition.

I approach this research objective in Chapters 3 and 4. Chapter 3 qualitatively explores the role of OSS activities of targets as an input to the strategic decision-making process leading to an acquisition. The chapter aims to create first evidence on if and how a target's OSS development influences it being acquired. It sheds light on OSS-related acquisition motives and if and how a target's OSS development activity influences an acquirer's target search, evaluation, and selection process. In doing so, this chapter aims to contribute to a better understanding of acquisition motives and the role of openness in the decision-making processes leading to acquisitions.

Chapter 4 aims to quantitatively create first evidence about the role of OSS activities for target search, selection, and evaluation processes. Specifically, this chapter aims to uncover differences between potential targets active in OSS development and potential targets not active in OSS development, differences within the group of firms active in OSS development, and lastly, the role of the acquirer's activity in OSS. In doing so, this chapter attempts to contribute to research on target characteristics influencing target search, selection, and evaluation processes, as well as acquirers' capabilities in those processes. Furthermore, it aims to contribute to research on the strategic dimensions of a firm's openness decision (Chesbrough, 2006; Dahlander and Gann, 2010). Specifically, it aims to provide first evidence about the role of a firm's openness—in this case, participation in OSS development—for its chances of getting acquired.

2.3.2 Evolvement of OSS Development Activities as an Outcome of Acquisitions

The second perspective on the connection between OSS development and acquisitions focuses on the outcomes of acquisitions. It takes a different perspective as I now conceptualize OSS development as a subject potentially influenced by acquisitions. Prior research has covered the impact of acquisitions on firm performance (e.g., Haleblian and Finkelstein, 1999; Larsson and Finkelstein, 1999; Meglio, 2009), proprietary innovation after an acquisition (e.g., Ernst and Vitt 2000; Hitt, Hoskisson, and Ireland 1990; Hitt, Hoskisson, Ireland, and Harrison,

1991; Kapoor and Lim 2007; Ranft and Lord 2002), and stakeholders, such as customers or competitors (Kato and Schoenberg, 2014; Rogan and Greve, 2015; Valentini, 2016). However, research has never looked at open innovation activities of firms, nor interactions in communities, when it comes to the impact of acquisitions. As formulated in Chapter 1, the second research objective of this thesis therefore is:

- **Research objective 2:** Create an understanding of the impact of an acquisition on the evolvement of a firm's OSS activities, the OSS projects it works on, and the related communities.

I approach this research objective in Chapter 5 in a mixed-methods study combining quantitative data analysis with qualitative insights from interviews. In an exploratory manner, this chapter aims to shed light on the impact of acquisitions on OSS development. It covers the impact of acquisitions on the target's activities in OSS projects as well as the impact on the community around them. Furthermore, contingencies across acquisitions and the drivers behind those contingencies will be covered. In doing so, this chapter aims to contribute to research on the evolution of OSS projects and factors influencing the evolution of the informal collaboration of firms and communities in such projects. Furthermore, it also aims to advance research on the impact of acquisitions on innovation and researchers by focusing on open innovation instead of proprietary innovation and covering the impact of acquisitions on external collaborators such as community members.

Qualitative Study: How Does the Involvement of Firms in OSS Matter in the Pre-acquisition Phase?

3

3.1 Introduction and Motivation

OSS development on public platforms has many characteristics that differ from traditional innovation processes (Raymond, 2001). Most distinguishable, the development takes place "in the open," which usually involves a community of contributors who voluntarily contribute to the development process, and the software is often available for free under the terms of an OSS license (Raymond, 2001). Both aspects are interesting for acquisition research. First, why should a company active in OSS get acquired? After all, at least parts of the target's code are often available for free on the web rendering an acquisition unnecessary if access to the companies' codebase is the goal. Second, the fact that the development takes place in the open potentially presents new opportunities for acquirers to evaluate a target based on its OSS activities. This is not possible in the same way for firms focusing on proprietary software development or other forms of proprietary innovation as those innovation processes, and their outputs are typically kept secret or at least protected. The question is: is the activity of firms in OSS actually relevant for acquirers in the acquisition process, and if so, how?

Prior research has not examined this question yet. Prior acquisition research has generated an understanding of general acquisition motives and studied some target, market, or acquirer characteristics relevant for target search, selection, and evaluation (e.g., Cunningham et al., 2020; Grimpe and Hussinger, 2008; Ouimet and Zarutskie 2012; Ransbotham and Mitra, 2010; Warner et al., 2006; Worek et al., 2018). However, it has not covered the role of targets' activities in an

Electronic supplementary material The online version of this chapter (https://doi.org/10.1007/978-3-658-35084-0_3) contains supplementary material, which is available to authorized users.

open innovation environment. OSS research, on the other hand, has generated a good understanding of the benefits and disadvantages of a firm's engagement in OSS (e.g., Ågerfalk and Fitzgerald, 2008; Bonaccorsi et al., 2006; Dahlander and Magnusson 2005; Gruber and Henkel, 2006; Henkel, 2004, Nagle, 2018a). However, it is not clear how those advantages and disadvantages translate into acquisition motives and how—if at all—such activities are considered by potential acquirers.

This chapter aims to address the above questions concerning the role of a target's engagement in OSS development for acquisitions. More specifically, it examines how a firm's being active in OSS influences the selection and evaluation of this firm as an acquisition target. This requires—first—to explore the reasons *why* firms active in OSS get acquired or not; and second, to understand *if and how* the engagement of the target in OSS, where development processes including the feedback from the contributor and user community are well observable, can influence target evaluation processes.

To create an initial understanding of the different aspects of the role of OSS in the target selection and evaluation process, I conducted 52 interviews with acquirers of targets active in OSS development and managers, employees, and community members of targets complemented with secondary data from press releases and the web. Interviewing acquirers, targets, and community members allows to holistically understand the role of OSS for the target selection and evaluation from different angles.

When analyzing the interviews, three clusters of topics regarding the role of potential targets' OSS activities in target search, selection, and evaluation processes emerged. Specifically, the interviews revealed OSS-specific acquisition motives, how OSS activities can help acquirers in their target search and evaluation processes, and how OSS licenses are an important factor to consider in acquisitions. These findings are relevant for research on acquisition motives, information asymmetry between target and acquirer, and license choice in markets for technology. As this study is not without limitations, I suggest to perform further (quantitative) research on this phenomenon at the end of this chapter.

I structure the remainder of this chapter as follows: first, I explain my motivation for starting this research project with a qualitative analysis and explain my sampling, data collection, and data analysis strategy. Next, I discuss results along the three clusters of findings that have emerged during the data analysis. I conclude with a summary of findings, contributions to research, managerial implications, limitations of the research, and suggestions for further research.

3.2   Method

3.2.1   Qualitative Study Design

The qualitative research study presented here aims to create an initial understanding of the different aspects of the role of OSS in the target selection and evaluation process. I aim to identify concepts, patterns, and critical parameters of the decision-making process behind acquisitions of firms active in OSS development, focusing on the role of OSS for those. Given the novelty of this research, qualitative research is a valuable starting point to develop theory about this under-theorized phenomenon based on real-world observations from acquisitions (compare, e.g., Siggelkow, 2007). It is also a good methodological fit given the absence of prior empirical evidence and literature dealing with acquisitions in the context of open innovation (Edmondson and McManus, 2007; Eisenhardt, 1989; Merriam, 2009). Qualitative research based on real cases is particularly suited to examine *how* and *why* questions (Eisenhardt, 1989; Eisenhardt and Graebner, 2007; Yin, 2003); the study at hand explores the motives behind the strategic decision to acquire a target active in OSS development and the influence of the target's OSS activities on the acquisition decision making. I select a multiple-case study design as they are typically more compelling, generalizable, and robust than single-case studies (Graebner and Eisenhardt, 2004). The chosen multiple-case study design constitutes a more substantial base for theory building, making the emergent theory better grounded (Yin, 2003) and allowing for a close correspondence of theory and data (Glaser and Strauss, 1967). I also use an embedded research design (i.e., multiple levels of analysis) that includes different perspectives from targets and acquirers on strategic (i.e., acquisition motives) and operational topics (i.e., the process leading to acquisitions), an approach well suited to generate richer, more reliable insights (Yin, 2003).

Qualitative research is becoming increasingly popular in the management field, which is reflected in an increasing number of scientific publications in high-quality journals (Bansal and Corley, 2011). Combined with qualitative research being highly regarded for providing profound insights and the proximity to the phenomenon (Bansal and Corley, 2011), I believe this method is the right choice for initially exploring the phenomenon of acquisitions of firms active in OSS development.

The study presented in this chapter is based on 52 interviews conducted between early 2019 to mid-2020, complemented by secondary data. The data collection effort described here was part of a broader study carried out to uncover the role of OSS before and after acquisitions. The findings from the interviews about the impact of acquisitions on OSS development are presented in Chapter 5. As the data was collected from the same interviews, and to avoid unnecessary duplication, I present the whole data collection effort in this chapter. The study presented in this chapter mainly builds on insights generated from interviews with managers and employees of acquirers and targets. Interviews with community members of the targets were predominantly used for the study presented in Chapter 5, as community members have little to no influence on the processes taking place between the acquirer and the target leading to the acquisition. However, as a few community members were highly knowledgeable about targets and their acquisitions, I selectively also use their insights in this study.

3.2.2   Sampling of Acquisitions and Interview Candidates

A solid sampling approach is required for qualitative research aiming to build theory (Eisenhardt, 1989). The interviewees selected for a research project need to be "excellent participants to obtain excellent data" (Bryant and Charmaz, 2007). Given this research's exploratory nature, I employed theoretical rather than random sampling for selecting acquisitions and interviewees, which is generally considered appropriate in research designed to build theory based on a novel phenomenon (Eisenhardt, 1989). Theoretical sampling is an iterative sampling approach and is regularly used in grounded theory research, where the analysis of the data guides the researcher to new elements for the sample to fulfill a specific theoretical purpose (Strauss, 1987).

The sampling approach utilized in this study follows three steps. In a first step, I selected relevant acquisitions. In a second step, I searched for and recruited suitable interview candidates to cover different stakeholders' perspectives on the focal acquisition. In a third step, based on an initial analysis of the data obtained from the first interviews, I selected additional acquisitions and/or interview candidates to further examine specific categories within the emerging theoretical findings.

In the first step, I selected acquisitions of interest using acquisition data from Crunchbase[1] and index.co, a website curating a list of acquisitions in the OSS industry.[2] I chose acquisitions based on how well they appeared to represent the

---

[1] I am grateful to Crunchbase for kindly providing an API key for their database and granting me the right, under the terms of an academic license, to publish aggregate information based on their data.

[2] https://index.co/market/open-source/acquisitions (last accessed 10/18/2020).

phenomenon of interest, including extreme or deviant cases, confirming and disconfirming cases, and typical cases (Kuzel, 1992; Patton, 1990; Pettigrew, 1990).
Typical cases are necessary to calibrate what might be considered normal or average, while confirming and disconfirming cases allow elaboration on the initial
analysis, seeking exceptions, and looking for variation. Deviant cases allow the
exploration of highly unusual manifestations of the researched phenomenon. Particularly, I focused on covering acquisitions of different target sizes from small
start-ups to the largest acquired OSS-active companies, acquisitions in different
geographic areas covering North and South America, Europe, Asia, and Australia,
and lastly, different types and levels of involvement of the target and the acquirer
in OSS development.

In the second step, I sourced interview candidates related to the selected acquisitions, aiming to recruit interview candidates from three groups of stakeholders
to get a comprehensive picture of the role of OSS in acquisitions. The first group
is the owners, managers, and employees of the target. The second group is the
owner, managers, and employees of the acquirer. The third group is the community members contributing to the same projects as the acquisition target. Within
the groups, I mainly aimed at recruiting two types of interview candidates: C-level
executives and/or founders of the target and the acquirer and highly active OSS
developers from the target and the community. I focused on those types because
of their experience and expert knowledge about acquisitions. C-level executives
and founders are deeply involved in the entire acquisition process, particularly in
the strategic decision-making. Hence, I expected them to be well informed about
the topic and able to provide deep insights into both, individual acquisitions and
patterns across the sample of acquisitions. Highly active OSS contributors from
targets and their community were selected as they are closest to the OSS development process of the target and—due to their typically central role in the OSS
projects of the target—most knowledgeable about the OSS projects and the dynamics of the informal collaboration between the firm and the community relevant
for this study. I sourced interview candidates from LinkedIn, GitHub, and the
respective target's or acquirer's website. Employees on GitHub were searched by
matching the email domains used by GitHub users for their contributions at the
time of the acquisition with company domains available on GitHub. The broad
range of acquisitions and stakeholders I covered with my interviews provided an
opportunity to explore acquisitions in the context of OSS with different organizational environments like company and community sizes and relationships between
acquirer and target, and the role of OSS in the context of acquisitions for different
stakeholders.

In a third step, I selectively recruited additional interview candidates either for acquisitions already covered in the sample or for additional acquisitions based on insights from the initial cases' analysis. I particularly added additional acquisitions that shared the same acquirer with initially selected cases to distinguish differences between different acquisitions with the same acquirer. I also focused on adding different acquirers covering a broader range of industries. Mainly, I added acquisitions with private equity investors as acquirers to examine the role of OSS for mostly financially oriented acquirers. In total, 25 individual acquisitions in the years 2014–2018 are covered in this study. Several interviewees from acquirers covered more than one acquisition.

In sum, I, together with Prof. Henkel, contacted roughly 550 interview candidates via email or LinkedIn. Each candidate was contacted with a personalized cover letter. Follow-up emails were sent once to each candidate who did not reply to the first message. To motivate more accurate responses and minimize social desirability bias, interviewees were guaranteed confidentiality. The effort to source interviewees for this study resulted in 52 interviews. Four candidates rejected our request, mentioning that they are not knowledgeable about the acquisition. One declined, referring to a confidentiality clause he signed. For an overview of the acquisitions covered by the interviews, see digital Appendix A. The size of the acquisition targets covered by the interviews ranged from three to roughly 10,000 employees, with most firms being in the Crunchbase size category of 11–50 employees. The average age of the targets was eight years at the time of the acquisition (median: six years). Sixteen of the targets had disclosed funding received from outside investors prior to the acquisition, with an average funding of \$26.7 million (median: \$15.5 million). Acquisition prices were only rarely disclosed: Only for four acquisitions I was able to find acquisition prices, the highest being the Red Hat acquisition at a price tag of ~ \$34 billion.

In qualitative research, there is no ideal number of acquisitions or interviews (Merriam, 2009). For example, Eisenhardt (1989) recommends four to ten detailed cases, while Yin (2003) suggests six to ten cases, and Gentles, Charles, Ploeg, and McKibbon (2015) mention that also more than 15 cases are not uncommon. Given these benchmarks, the number of 29 acquisitions covered in 52 interviews in this study appears to be a lot. However, theoretical sampling which was used in this sampling, is rather concerned with theoretical saturation than requiring a specific sample size. Here, adding additional acquisitions and interview partners to the sample stops when the marginal utility of an additional acquisition or interview reaches zero—i.e. when theoretical insights saturate (Charmaz, 2006; Strauss and Corbin, 1998). Given the large variations of cases regarding the firms involved on target and acquirer side (e.g., firm size, relative size, industry focus) and different roles these firms played in OSS development before the acquisitions, a larger sample was chosen, and theoretical saturation was the stopping criterion in this study.

3.2.3 Data Collection and Analysis

This study's primary data source is 52 interviews, complemented by a secondary data review of more than 200 acquisition-related press releases and news articles from tech journals and websites. For the interviews, I developed—guided by extant literature on OSS and acquisitions—a semi-structured interview guide to provide a consistent structure and line of investigation for the interviews while also allowing for the discovery of emergent themes (Yin, 2003). Its open-ended questions with flexible sub-questions addressed different topics along the acquisition process (see digital Appendix B for the interview guide). Each interview covered the following topics: The role and relation to the firms involved in the acquisition of the interviewee; description of the acquiring and target firm with a particular focus on their involvement in OSS development; description of the acquisition motives, technology and market relatedness between the acquirer and the target, and potential prior collaboration between them; how OSS played a role in the process leading to the acquisition; and various aspects of the impact of the acquisition on the OSS development after the acquisition differentiating between different stakeholders, particularly target employees and community members. The interview guide for qualitative research should enable a "conversational mode" and an individualized conversation with each participant (Yin, 2015) so that complexity of the matter can be explored more deeply and the interviewee's point of view can unfold. Therefore, research on each acquisition and each interviewer was conducted before the interview, and the guide slightly adjusted to ensure that questions better reflected the interviewee's previous experience. Particularly, adjustments were made to capture the different views of managers and employees from the acquirers or the target and the target's community members. Lastly, slight adjustments to the guide were made over time to explore newly emerging themes.

Most of the interviews were conducted via web meetings, while two were conducted in person in Germany. Detailed notes were taken during the interviews, and all interviews were conducted in English or German. After receiving permission from interviewees, the interviews were recorded. The length of the interviews varied from 20 to 90 minutes. In most interviews I was joined by a second researcher, which Eisenhardt (1989) recommended as it allows to capture more insights during the interview process and enhance the creative potential of a qualitative study.[3] In interviews with two interviewers, one interviewer was responsible for guiding the interview partner through the interview guideline, and the other interviewer took notes and asked additional questions based on the interview partner's answers (Miles & Huberman, 1994). I recorded and transcribed 48 interviews filling roughly 1200 pages of double-spaced text, and took notes in four cases. The interviews were then coded using MaxQDA. Follow–up questions were asked via email when clarification was required.

Building on the approach of Gioia, Corley, and Hamilton (2013), I analyzed the interviews. An initial, yet fluid structure of categories for the coding was created from first theoretical considerations on the different topics covered in the questionnaire. During the coding process, iterations led to the adjustments of individual categories (Mayring, 2014). Three clusters of categories emerged: the acquisition motives of firms active in OSS (3.3.1), the role of OSS in the process leading to the acquisition, i.e., the target search, selection, and evaluation (3.3.2), and the role of licenses in OSS acquisitions (3.3.3). Within each cluster, I aggregated findings across cases and analyzed differences across cases.

Throughout the data collection and analysis, I took steps to minimize interviewee bias, which is a concern in interview-based qualitative research projects (Golden, 1992; Huber and Power, 1985). Whenever possible, I interviewed multiple individuals from the target, the acquirer, and the community. Those different interviewees should have different perspectives on the focal acquisition. I did not find significant differences in their answers, which should have been the case if bias were an issue (Seidler, 1974). However, it has to be noted that I was only able to recruit multiple interviewees for 13 out of 25 acquisitions, so results need to be understood with this caveat in mind. Furthermore, next to interviews, various additional sources from the web (e.g., press releases or articles related to the acquisitions on tech-magazines like techcrunch.com or wired.com) were reviewed to enhance the overall validity of the findings (Golden, 1992).

The approach for the data analysis described in this section allowed to create a holistic overview of the phenomenon of acquisitions of firms active in OSS development. I was able to derive coherent insights. The discovery of a variety of

---

[3] I conducted all interviews myself, for some I was joined by Prof. Joachim Henkel or either Katharina Beck or Johannes Huber, two master students of mine. I gratefully acknowledge their help in contacting interview candidates and transcribing the interviews they have joined and which they used for their own theses focusing on the role of individuals in acquisitions related to OSS development.

nuances regarding the role of OSS in the decision-making shows the phenomenon's richness and that a sufficient number of interviewees capturing the relevant perspectives were interviewed in the process.

3.3   Results

In this section, I present the main findings on the role of OSS in target selection and evaluation. During the interviews and the subsequent coding process, three clusters of themes emerged, where OSS plays a role in the context of acquisitions: first, and from a strategic perspective, the motives for acquirers active in OSS. Second, and from a process perspective, the role the OSS development of the target plays in the processes leading to the acquisition (mainly target search, selection, and evaluation). As a third cluster, the role of OSS licenses emerged, which touches both strategic and procedural elements. An overview of exemplary codes per category can be found in digital Appendix C.

3.3.1   Strategic View: OSS-related Acquisition Motives

Acquisitions are highly strategic decisions motivated by complex objectives and usually involving multiple acquisition motives (Worek et al., 2018). Typical motives include access to (complementary) resources or new technologies, exploiting economies of scale and scope, geographic expansion or the expansion of the product portfolio, acqui-hires (i.e., acquisitions of the team and their expertise), acquisitions of competitors to pre-empt competition, and acquisitions motivated for financial reasons (see e.g., Chatterjee, 1986; Cunningham et al., 2020; Grimpe and Hussinger, 2008; Hitt, Hoskisson, Johnson, and Moesel, 1996; Ouimet and Zarutskie, 2012; Seth, 1990; Walter and Barney, 1990; Worek et al., 2018). I found all of those motives in the acquisitions covered in this study. Across all acquisitions, three motives were prominent: acqui-hires (12/25 acquisitions), complementary product acquisitions (9/25), and technology acquisitions (7/25), while other themes, like acquisitions focussing on financial aspects or acquisitions of competitors to pre-empt competition, played a minor role (2/25 each). Most importantly, the interviews showed that OSS-related acquisition motives exist—firms get acquired because of their OSS development activity. In fact, interviewees mentioned OSS-related acquisition motives for 18 out of 25 acquisitions, showing that firms active in OSS often get acquired for—among others—their OSS development activities and the communities they have built. The interviews also revealed that different OSS-specific motives to acquire firms active in OSS development exist and how those relate to the known motives from the literature. Figure 3.1 shows my proposed extension of known acquisition motives with OSS-related acquisition motives based on the coding of the interviews.

Code category: Motives

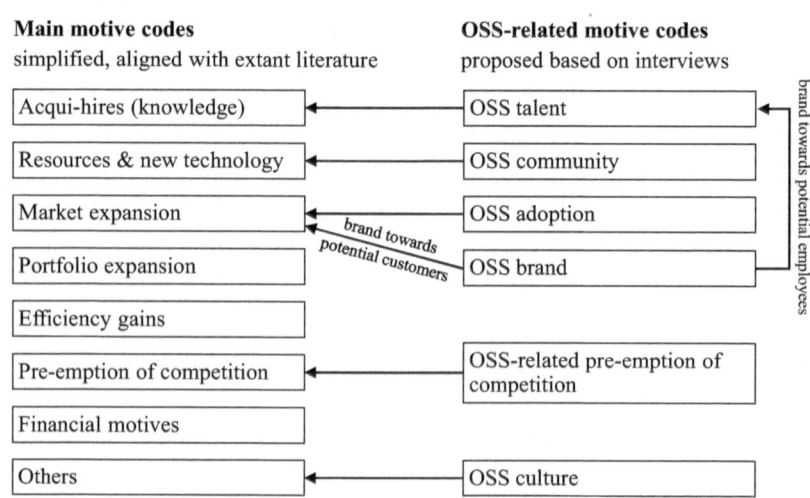

Figure 3.1   Acquisition motive coding and proposed extension of known acquisition motives with OSS-related motives

Not all acquisitions were motivated by OSS-related motives; also, some acquisitions had other motives next to OSS-related motives. Those were mostly acquisitions of a complementary technology or a complementary product offering or acqui-hires. I do not go into the details of non-OSS related motives here as they are not at the focus of this study. Instead, I focus in the following on the different OSS-related acquisition motives I identified (useful information on technology acquisitions and motives for such in general can be found, e.g., in Stein, 2017).

OSS-related acquisition motives
OSS talent. Developing OSS requires specific skills like being knowledgeable on how to contribute to OSS projects or how to manage an OSS community. Interviews highlighted that these skills are rare and can constitute a competitive

advantage. As a result, acquirers buy targets active in OSS because of these specific skills of the target's employees—an OSS talent acquisition. A chief architect who has been part of OSS-related acquisitions as a manager at an acquirer and a target explained how OSS talent make for an attractive target: *"When you acquire an Open Source heavy company you get a set of folks who are fluent in the technology stack and interacting with other parties [...] and working in the open. That's kind of hard to find if you haven't done it."* Often, the acquirer not only acquires the OSS-related skills of the employees, the acquirer also wants to learn those skills from the target as one software engineer of a target explained: *"[The acquirer said] we want you to stay like you are because you are doing things in a way we can't do. You need to teach us these things."*

Acquirers seem to see the acquisition of OSS talent as a way to create a potential competitive advantage. The Chief Executive Officer (CEO) of a target explained: *"No potential acquirer seemed to make the point 'Open Source, we have to be careful'. It was more like if you have expertise and differentiate [from competition] in this field, that was presented as something special and was actually seen as an advantage."* The CEO of an acquirer expressed the same view, explaining how acquiring a target central to an OSS project can create a competitive advantage: *"I think when you acquire the company around the project, you don't really acquire a project, that's difficult to do, but you acquire the company; that company usually has in its staff, the people who may have started the project, some of the best engineers that are participating in content contribution to the project. So, I think there are two competitive aspects: One is talent acquisition in one motion, the other one that you are now using that acquisition of mindshare as a competitive bullet. [...] you go out to the broader community of customers and you say: 'Look, we acquired the company behind XYZ project, our competition doesn't have that; it creates an immediate emotional gap for the customer. The customer is looking at this and says: 'Oh, you know, why would I want to work with their competition, when all the people who built this and the mindshare now lives within the acquiring company?'"*

Acquiring firms active in OSS development for their specific skills and knowledge is related to acqui-hires. A chief architect of a target set OSS-related motives in the context of acqui-hires: *"In many ways, I think the acquisition by and large was a talent acquisition and focused both on the assets that we did have and [OSS project]."* These acquisitions are highly prominent among start-ups and in the technology sector (Chatterji and Patro, 2014). Hence, I propose that acquiring to gain OSS talent should be considered a special case of an acqui-hire.

OSS community. Having access to an OSS community can constitute an external resource for a firm (Dahlander and Wallin, 2006). The interviews revealed that acquirers buy targets for this particular resource. One chief architect of a target explained how the acquirer, itself active in other OSS projects, sought access to the community: *"Our position in the community, and Open Source ecosystem, was important because [acquirer] is making strategic investments in [OSS-heavy technology]. […] So, I think with respect to the [target's] community, which is where [acquirer] wanted to go, it was highly important for them to bring in people who had experience in the community and had a good reputation."* The same motive was confirmed for an acquirer not itself active in OSS by a platform developer of a target: *"[Acquirer] was before a kind of closed company. […] And so they wanted to kind of learn how to also become a part of this community. […] They wanted to learn how to get more traction with that community."*

An alternative way for acquirers to buy access to a community would be to organically build access by joining the community as a contributor or build a community from scratch. However, acquiring a company that has already successfully built a community can be a faster, less risky, and less costly way to do so. As one CEO of an acquirer explained: *"So, if you were to think of the cost and energy required to build a community […] from scratch it requires more than just money. It requires the right personalities involved, the anchor people that might be the focal point of a project that get people excited, energy and money, time to invest in community building. It's something that I think for most companies especially those who aren't experienced in building communities find it either impossible or so expensive that they have no desire to approach the problem; and the expensive isn't in just building, but it's in the risk of failure. They can put in all this money and energy and they don't succeed and now it was a giant waste of everybody's time. So, I think one of the primary motivators in Open Source acquisitions is that it's actually the least risky path [to get a community]."* The example also highlights the role of key developers in OSS communities. Gaining access to communities requires control of the key developers and community managers in a community, a notion I came across in several interviews. Lastly, the CEO also explained that the size of the community can matter for an acquisition: *"The bigger, the better—primarily because that's the things that is irreproducible by the acquirer. They would have a very hard time building a competing community or overtaking the community in some way."*

OSS adoption. Given the open, and often free nature of OSS, OSS is often adopted faster than proprietary software. Firms achieving high user adoption through OSS seem to be attractive targets. When asked if the OSS development of the target played a role when they decided to buy a target, the Chief Technology Officer (CTO) and co-founder of an acquirer explained: *"With [target], very much so […] we wanted something that had a wide adoption. And the only way to get wide adoption in that space was to be Open Source."* A target's employee confirmed this acquisition motive and explained in more detail how OSS drives user adoption: *"I think a lot of times companies do want to acquire the user, [...] and I think with [target] had their Open Source [...] and it's used by lots of lots of projects. [...] that can make a company very attractive. And sometimes Open Source projects will get adopted quickly, because they're Open Source and I think people are attracted to sort of the 'freeness of speech'-part of Open Source. But I think there are also a lot of times when people are attracted to the free [part of Open Source]. I think that helped drive [target's] adoption a lot."* Lastly, the CEO of an acquirer explained why being active in OSS can make targets in the technology sector attractive for acquirers: *"The risk in [in this context the challenge for success for] start-ups or technology businesses is not technology—it's customer acquisition. [...The target's OSS project] bringing in a huge amount of traffic allowed us to convert a very, very small percentage of them into customers and still build a decent business."* Interestingly, he added that after the acquisition, the acquirer was open-sourcing parts of their software base *"to help with this customer acquisition strategy."* Thus, the acquirer not only acquired the large user base of the target but also adopted the target's OSS-based user acquisition strategy.

OSS culture. OSS development is highly connected to a unique culture centred around the users of the software (O'Mahony, 2003; von Hippel and von Krogh, 2003). This culture seems to be in demand at companies nowadays. Therefore, the target's familiarity with the OSS culture can make for an attractive target in acquisitions. A target's platform developer explained the acquisition motive: *"I think they kind of felt that they were falling like culturally behind at that times. And so [...] they were looking for not only kind of the technology but also kind of like the culture of software development."* Another acquired senior principal engineer explained that the acquirer's goal was that the acquirer would adopt the target's culture: *"They [acquirer] recognized that their culture was not going to take them further. [...] they wanted [acquirer] to become more like [target]."* The same motive, framed in the reverse direction, was also mentioned by the CEO of an acquirer active in OSS: *"We purposely avoided those start-ups that had a closed model because we knew that would be really challenging to turn [those] in an Open Source and try to drive the [OSS] standards if there wasn't already kind of an Open Source ethos or Open Source mindset in the company."* Hence, the acquirer already active in OSS development avoided targets that were not Open Source to avoid a cultural misfit.

OSS brand. Open Source development is known to be considered attractive by many developers. Companies can build a brand in the broader OSS community and among (potential) customers by contributing to OSS. This notion of OSS can also be found among acquisition motives: Firms with no or little presence in OSS acquire firms active in OSS to strengthen their brand. A senior software engineer explained: *"I don't think [acquirer] had any Open Source. And with the addition [of target] to the portfolio, of course, now at least I think 150 more repositories that are Open Source. This is a representation in the Open Source world for the company; you can represent yourself and build a profile for yourself."* Given the regularly large number of users, acquiring a company with a large user base of their OSS can be a comparatively cheap way to build a brand as the CEO of an acquirer explained: *"Customer acquisition is always a challenge in any software business. And often very expensive. So, we were excited about the possibility of brand awareness and organic traffic that [target's OSS project] would bring."* Not only the number of users is valuable for building a brand in OSS, but also the community's behavior—namely word of mouth marketing—can contribute to building a brand by acquiring an OSS-active company. A target's senior principal developer explained: *"The community is valuable [for the acquirer] because, for one thing, it's a lot of interpersonal level marketing, right? So, something that has a strong community will also tend to have a lot of word of mouth. Word of mouth marketing, right? Something that people want to use that they're so passionate about using and improving; that they're gonna join an Open Source community—and participate in a community."* While most interviewees mentioned OSS brand as an acquisition motive in the context of brand towards customers, the last example also shows how building a brand through an acquisition of an OSS-active company can also be relevant for building a brand in the developer community. This may be relevant for hiring talent[4] and to increase the number of external contributors to a project, i.e., increase the availability of the community as an external resource to the acquirer.

OSS-related pre-emption of competition. Firms may engage in acquisitions to pre-empt competition in technology markets (Grimpe and Hussinger, 2008). Two mechanisms of how pre-emption of competition can be a motive to acquire a firm active in OSS development were found during the interviewing process. However, each was only mentioned by one interviewee, and in one of the cases, another interviewee mentioned diverging motives for the acquisition at hand. In the first case, a community member suspected that the acquirer shut down the OSS project because they wanted to use the tool internally and exclude competition from using the tool. When asked why the acquirer stopped the project, he said: *"And my suspicion is [the acquirer] wanted that for themselves. Maybe they are using it*

---

[4] See Lerner and Tirole (2002) or Spaeth et al. (2015) for how firms can present themselves as an attractive employer by letting their employees contribute to OSS during working hours.

*in their cloud; who knows, I suspect that is what happened. But that is why I think
[target] was valuable to them. And if it is Open Source, it is available to their competitors, too, you know.*" This notion of pre-emption of competition is related to
the "buy before someone else does"-acquisition motive described by Fridolfsson
and Stennek (2005), where an acquirer buys a firm to prevent the firm getting
acquired by another competitor—or in the case of OSS, the OSS being used by
another competitor. In the second case, another pre-emptive motive—namely a
so-called "killer acquisition" (Cunningham et al., 2020)—was suspected. A killer acquisition in the context of OSS development is the acquisition of the target
to shut down the target's OSS project, which poses a competitive threat for the
acquirer's own OSS project or their proprietary product. The target employee in
the case at hand stated: "*[…] it was basically an acqui-hire of the team. Mainly. So,
there was that, and some combination of [acqui-hire with] them seeing [target] as a
threat in some way.*" However, the CEO of the (OSS-active) acquirer of this target
denied the motive of shutting down a competing OSS project: "*We were always
acquiring for engineering talent and or product augmentation.*" This last example
of OSS-related motives is a reminder that even though rigorous execution of the
qualitative data collection and analysis was conducted, results might be biased as
interviewees might be unwilling to reveal to the interviewers (and possibly even
themselves) the real motives behind acquisitions, or might only have conjectures
regarding these motives.

Due to the very limited presence of the "pre-emption of competition-motive"
in the interviews and no other such cases found in my web research except the
infamous MySQL-Oracle acquisition, one could suspect these have comparatively
little prevalence among the population of OSS-related acquisitions.

Result 1. OSS-related acquisition motives exist. I propose to differentiate six
different OSS-related motives: OSS talent, OSS community, OSS adoption, OSS
brand, OSS culture, and OSS-related pre-emption of competition. Many firms
active in OSS development get acquired for OSS-related motives, but not all
acquisitions of such firms are related to such motives.

Differences across acquisitions
Not only did the cases reveal different motives related to the OSS activity of
targets, but also differences across acquisitions and their motivation. Motives seem
to vary depending on the level of the OSS activity of the main parties involved in
an acquisition—the target and the acquirer—before the acquisition.[5] Based on the

---

[5] I did not find specific patterns regarding firm size of the target or acquirer by the number of
employees. I did also not find specific differences in acquisition motives regarding geographic

findings from the interviews, I identified four combinations of the level of OSS activity of the target and the acquirer that lead to different patterns of acquisition motives (Figure 3.2; see digital Appendix D for an assignment of the cases to the boxes in the figure).[6]

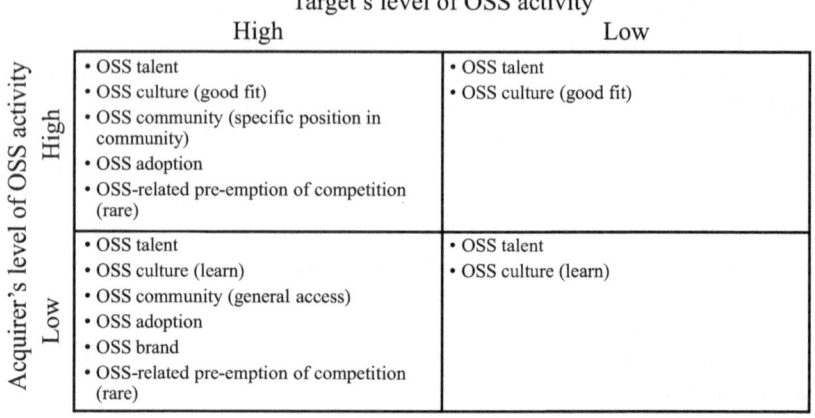

Figure 3.2   OSS-related acquisition motives and target's and acquirer's level of OSS activity

High level of OSS activity of target and acquirer. In this quadrant, both firms are active in and accustomed to working in the OSS environment. In some cases, they might even have collaborated before the acquisition. Many motives have been identified for this quadrant. OSS talent has been mentioned for almost all cases—OSS active acquirers seem to acquire other OSS firms because they have resources that have already experience with working in OSS and can hence (more easily) work on different projects across the combined entity. Furthermore, and different from other quadrants, acquirers acquire OSS targets for their specific position in a specific community; the acquirer wants to broaden their presence in

---

regions, yet, this finding has to be understood with the caveat in mind that many acquisitions in my sample took place in the US.

[6] The level of OSS activity needs to be understood on a relative scale: Large corporations with very little activity are considered a "non-OSS-active" firm as they are inherently not OSS oriented (for example almost all firms with more than 100.000 employees have *someone* in their organization contributing on GitHub, but that doesn't make almost any of them an OSS oriented firms).

OSS by joining a community they have so far not been present in and acquire a key player in this community. Acquirers active in OSS valued the good cultural fit of targets that have already been active in OSS, a pattern shared by all acquisitions, where the acquirer was highly active in OSS. Another prominent motive was acquiring the large user base of targets that have successfully built a large community around them (OSS adoption motive), a motive shared by all acquisitions where the target was highly active in OSS development. Furthermore, one case was found where the acquirer might have acquired the target to stop an emerging OSS project, potentially threatening the position of its own OSS project.

High level of OSS activity of target and low level of OSS activity of acquirer. In this quadrant, the targets have a high level of OSS activity, follow an OSS-based business model, and/or have successfully built a large community around them, while the acquirer has little to no experience with contributing to OSS. Compared to the first quadrant, the acquisition in this quadrant is characterized by the acquirer's wish to gain general access to OSS development, the OSS ecosystem, and the OSS culture. First, they seek access to communities, where it seems that the acquirers are less specific about the particular community they want to join. Instead, they particularly value if the target successfully built a community in general—a capability the acquirer does not possess so far. Second, the acquirers seek access to the "OSS culture." They acquire OSS-active targets to get higher adoption of the OSS culture within the acquirer—at the risk of creating cultural misfit within the combined entity. This is a key difference to OSS-active acquirers, who buy OSS-active targets to avoid misfit. And third, acquirers buy targets active in OSS to build a brand in the OSS ecosystem to attract customers and talent— which is again different to the first quadrant, where the acquirers typically are "a brand" within the OSS ecosystem already before the acquisition. In line with these motives, the OSS talent-motive was also often mentioned in this quadrant. Lastly, even if they seem to happen rarely, acquirers focusing on proprietary software development might acquire targets active in OSS development if they are essential in an OSS project that is a potential (future) competitor to their proprietary software product. The MySQL-Oracle acquisition in 2010 is considered an example of such an acquisition.

Low level of OSS activity of target and high level of OSS activity of acquirer. In this quadrant, targets have a comparatively low OSS activity level, while acquirers are frequently contributing to OSS. In these acquisitions, the community-related motive was less prominent. Instead, the OSS activities of the target were interesting for the acquirer from a competence perspective (OSS talent motive) or culture perspective (OSS culture motive), i.e., for being able and accustomed to working in the open environment and being used to participate in communities. Furthermore, for such targets, traditional, non-OSS-related acquisition motives were more prominent. This finding is not surprising: it can be assumed that for firms allocating many resources to OSS development, OSS development is closer to their core business than for firms allocating fewer resources to OSS. Firms with a high level of OSS activities will then be acquired for their specific OSS-related activities, i.e., the projects and communities they have built. In contrast, firms allocating many resources to other activities besides OSS development will be acquired for their non-OSS-related activities resulting in non-OSS-related acquisition motives or their general ability to work in an OSS environment.

Low level of OSS activity of target and low level of OSS activity of acquirer. In this quadrant, both the target and the acquirer have a comparatively low level of OSS activity. Acquisition motives are very similar to the quadrant mentioned above. Non-OSS-related acquisition motives are more prominent, and if the acquirer values the OSS activities of the target, the acquirer is particularly interested in the talent or to learn the "OSS culture" from the target.

Result 2. There are differences in the prominence of OSS-related motives depending on the target's and the acquirer's level of OSS activity. Most importantly, OSS-related motives are more prominent and more diverse for targets highly active in OSS development. Regarding differences across acquirers, OSS-active acquirers acquire targets active in OSS because of expected good cultural fit or to get access to specific OSS talent or a leading position in specific communities, while acquirers not active in OSS acquire targets active in OSS to learn the OSS culture or build a brand in the OSS ecosystem.

3.3.2   Process View: Roles of OSS in the Process Leading to the Acquisition

The decision-making process leading towards acquisitions is highly complex. Specifically, it involves significant uncertainties and risks associated with the target (see Section 2.2.4). During the interviews, it became clear that the target's OSS development can play a role in the decision-making process of the target beyond OSS-related acquisition motives. Specifically, OSS can play a role in *target sourcing* and *target evaluation*. Again, differences across acquisitions were found, which are clarified at the end of this section.

Two roles of OSS in the process leading to the acquisition

OSS and target sourcing. Target sourcing is the process of getting access to potential targets. Interviewees highlighted that OSS can be a *channel for target sourcing.* The co-founder and corporate development leader of an acquirer explained when asked if OSS can help to find potential targets: *"You know, I maintain a spreadsheet of interesting companies, and one of the other tabs is interesting Open Source projects and their relationship to companies. So, yes."* A senior manager of an acquirer mentioned that searching for targets in OSS communities is particularly suitable to get access to emerging technologies: *"I mean, had they not been doing it [target search] in Open Source, I don't think that acquisition would have ever happened. [...] The technology was so early. [...] So, it was really like, where do you find experts in this tech? You go to the communities that are building them, and so I think that was it [how acquirer found the target]."* Similarly, when discussing the role of OSS for acquisitions of start-ups, a senior principal engineer explained that operating in OSS can make start-ups visible for acquirers: *"It's free publicity, right? And it's discoverability too."* Another way for acquirers to find suitable targets is through direct collaboration in OSS projects, which was highlighted by a target's senior software engineer: *"Someone in [target] was working on a common [project] with [acquirer]. So, that's what actually got us in touch with [acquirer]."* The CEO of an acquirer mentioned how his company purposefully set out to collaborate with targets in OSS to get to know them better as potential targets: *"We knew we had to provide [specific technology] for our customers; there were three or four or five different start-ups in and around the [specific technology] space, and they were all Open Source start-ups. And so that was the context within which we were collaborating with them and could see the quality of their work and team."*

Result 3. Being active in OSS makes firms more visible towards potential acquirers.

OSS and target evaluation. Evaluating targets is difficult given that full information about the potential target, its employees and their skills, its technology, and its customers is not available. The interviews highlighted that OSS can be a useful *source of information* in target evaluation for acquirers, reducing risk in the decision-making process. A target's senior principal engineer explained different factors that make firms active in OSS easier to evaluate as potential targets: *"The difference between a company that is a start-up that embraces Open Source, and a company that doesn't, is [... the latter] give you next to zero information, right? When you can see the code, obviously, you can see the technology, but you can [also] see what decision-making process goes into there. And it allows you to, perhaps infer who are their customers; how many customers do they have? What kind of scale are they really operating at that guides their prioritization of technology problems? And you can learn a lot about what the context is, by reading sort of the tea leaves that you have access to in Open Source."* First, he emphasized the prospect of better assessing the target's employees and their decision-making processes; second, he highlighted the possibility of assessing the target's technology better; and third, he underlined the opportunity of assessing the target's customer base better. All three factors are essential for target evaluation as they can help reduce technology, market, and integration risk associated with acquisitions and have been mentioned across several interviews. Figure 3.3 shows which information about the target's OSS activities can be valuable for target evaluation, and how they are associated with risk reduction.

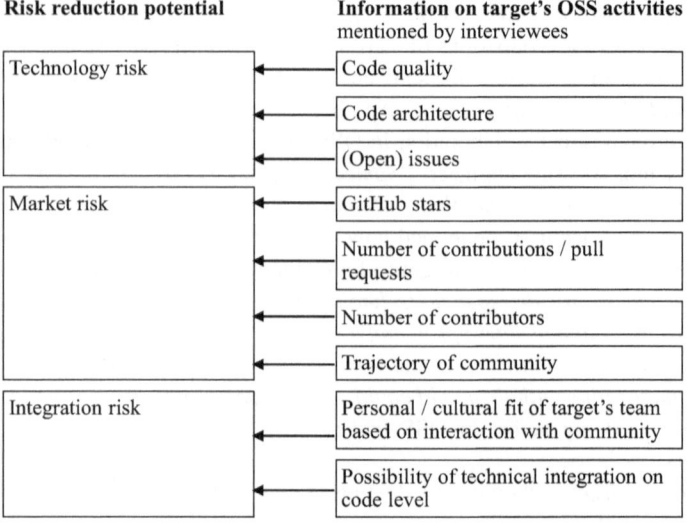

Figure 3.3   Information on target's OSS activities and risk reduction potential

Assessing the target's technology is key, particularly in technology acquisitions. The CEO of an acquirer explained how evaluating a target's OSS activities can help to de-risk decision-making in acquisitions by reducing the technology risk: *"the reason Open Source is interesting is that you can, unlike acquiring a proprietary company or a company with proprietary software, you can see the quality of the engineers before you buy, because you can see their Open Source contributions, you can go through their repos and see how they think about products and architecture. So, in our point of view, it can really help de-risk an acquisition because so much of the IP is accessible before you actually buy."* Another CEO of an acquirer noted how they check different information on GitHub to understand potential technological risks associated with the target: *"I have advised some acquisitions and one of the things we'll do is do a proper code review and architecture review of the technology, review outstanding issues in GitHub, for example, to look for any major bugs or security risks that may turn into extended or unexpected cost for the acquirer."*

Understanding the target's OSS activities can also help to reduce market risk in acquisitions. Particularly, acquirers can get an idea about how many customers or community members a target has and how they perceive the target. A co-founder and CTO of an acquirer explained how they checked the number of stars, a measure of project popularity in GitHub,[7] of a target's project during the acquisition process. He further elaborated how they checked community-related metrics: *"We look at the Open Source adoption; we look at the community, the Open Source community they built around the product to see what that looks like; We look at their contributions, how many contributions they had; all that sort of stuff."* A CEO of an acquirer explained how the development of the community over the last months before the acquisition is an important factor during the due diligence: *"In the due diligence, for example, imagine a community where the commit rate has precipitously dropped over the last nine months, or the number of contributors and their activity goes down, there's a question in the acquirer's mind—or there should be—of the validity of the community motive for buying the company. If it's increasing and it's increasing at a rate that is higher than linear, then there's something to be said for the value increase of the community."* Regarding the possibility of using OSS as an information source to reduce market risk in acquisitions, it has to be noted that this source of information is particularly useful to understand user adoption of the target's technology. Insights on the risks related to the commercialization of the technology cannot be derived from the OSS activities—an important factor to consider given that commercialization of OSS requires specific strategies to be successful (see e.g., Raymond, 2001).

---

[7] Stars are considered an indicator for project popularity in GitHub and have previously been used in computer science research (e.g., Papamichail, Diamantopoulos, and Symeonidis, 2016). Sanatinia and Noubir (2016) describe them as the "equivalent of a 'like' in social platforms such as Facebook." According to Borges and Valente (2018), GitHub stars function mainly as a tool for GitHub users to bookmark projects they use, bookmark projects they find interesting, and to show appreciation to the creators of the project.

Lastly, gaining a better understanding of the target's employees and their culture can reduce integration risk. The CEO of an acquirer explained how prior collaboration can reduce integration risk by ensuring cultural fit: *"The problem that you have if you don't do that [collaborate in OSS] is you might have cultural clash; the acquisition may not integrate well. So, by working alongside people, you get a chance to work with them and understand what they're like in practice."* He further elaborated on how collaboration in an OSS project between target and acquirer helped to better assess the target, on a personal as well as technical level: *"This is the beauty of the Open Source collaboration. […] We were developing interfaces so networking providers could plug into [acquirer's OSS technology]. And [target] was already making pull requests against those interfaces, and inevitably, in those pull requests, there's references to their product, which allows our engineers to then go look in their repos and see what is going on. […] The nature of pull requests in Open Source collaboration gives you first-hand exposure to, I guess, to two levels that are really important, right? The code and IP itself, but then you also get exposure to the engineers and the humans. You see the quality of people that the potential target has, how they think about technology, how they think about their company and their products. And so, you are really getting an assessment, not just of tech, but you're getting an assessment of humans as well."* A target's senior principal engineer joined the notion that an acquirer can gain insights about a target employees' personality from pull-requests: *"Where I think you can make very meaningful observations are interpersonal skills. […] How well did a person [from potential target] engage with the community making that pull request? How well did they accept feedback?"* Besides cultural fit, evaluation of the target's OSS activities can also help to test the possibilities of technological integration of the target's and the acquirer's code. The co-founder and CTO of an acquirer explained: *"What it [target's OSS project] allowed us to do was actually test out the integration ourselves before even talking to them. […] Because […] once you start negotiating, everything becomes weird. It let us do a lot of work on our own so that we could really understand what was going on."*

Result 4. Evaluating different OSS activities of a target and the code written by its employees, how it interacts with the OSS community and how the OSS community reacts to the target can help acquirers to reduce technology, market, and integration risk in acquisitions.

Differences across acquisitions

The examples above show that OSS can play a role in target sourcing and target evaluation. However, they have not been present in all acquisitions. Their relevance is contingent on the acquirer's abilities to understand OSS development and the amount of OSS activity of the target. As one co-founder and CTO of an acquirer explained: *"I think you have to understand the Open Source community and understand how to have value in communities. I think that's what it comes down to."* Accordingly, the aspects of OSS mentioned above have only been mentioned for acquisitions where the acquirer was active in OSS itself. This finding is in line with prior research, which found that technological overlap—in our case, target and acquirer being active in OSS—helps the acquirer to better evaluate the target in technology acquisitions (Hlavka, 2019). Furthermore, the aspects of OSS mentioned in this section are particularly relevant when the target is highly active in OSS, as especially in these cases, it is valuable to examine the projects and the interactions of the community with the potential target to understand the target better. For targets with lower levels of OSS activity, only direct collaboration in OSS projects between an acquirer and a potential target seems to be a relevant aspect of OSS development for target search and evaluation. Hence, these modes do not replace traditional ways of sourcing or evaluating targets but rather complement them. Regarding target sourcing, for example, many traditional methods were mentioned throughout the interviews; particularly, contact between target and acquirer was often made via supplier-customer relationships, investors, personal connections, and/or networking events.

Result 5. Evaluating a target's activities in OSS is more important the more the target is active in OSS. Furthermore, the acquirer needs to have specific skills to do so.

3.3.3 The Role of Licenses in OSS Acquisitions

A key learning from the qualitative study was the importance of licenses in OSS-related acquisitions—a topic I did not cover in my initial interview guideline. One CEO of an acquirer mentioned when asked about other factors relevant in OSS-related acquisitions not covered by the interview so far: *"You have not touched on licenses. [...] But I guess I'd encourage you to not underweight that in your studies because—it sounds obvious—but the GPL [General Public License; copyleft type] versus ASL [Apache Software License; permissive type] versus a free commercial license all have radically different impacts on how the companies integrate; on how the community continues to contribute; on the flexibility you have for monetizing what you purchase. And so, one of our criteria in an acquisition is understanding the Open Source license that the acquisition [target] is operating under."*

During this and the following interviews, it became clear that licenses used by potential targets play a role on a strategic level and a process level. On a *strategic level,* acquirers who want to integrate the acquired software with their own software seem to prefer targets associated with permissive licensed projects over targets associated with copyleft licensed projects. They do so because permissive licensed software can more easily be integrated in other Open Source or proprietary software, while copyleft licensed software requires all software it integrates with also to be copyleft licensed. The CEO explained how they avoid targets associated with copyleft licensed projects: *"The GPL is toxic, really, because everything it touches becomes GPL. And so that for us, when we are looking at [potential target's licenses], it comes up early in the process to be really clear; like one of the first things we do when we look at three or four companies to acquire. Well, the first thing we do is we compare the licensing models of each because we know that if we are going to get a GPL, it is going to create problems for us. So, it's early on in the process."* Another strategically important factor to consider when acquiring targets associated with copyleft vs. permissive licensed projects is the potential to commercialize the respective software. However, interviewees did not show a clear preference here: One interviewee highlighted the better options to monetize copyleft licensed software, as potential customers cannot use it in their proprietary software, which might increase their willingness to pay for products around the OSS software. Others highlighted the better option for permissive licensed software to continue it as purely proprietary software, giving the acquirer the option to sell the software without operating an OSS business model anymore.

On a *process level,* it became clear that if an acquirer considers acquiring a target associated with copyleft licensed projects, technical due diligence and integration planning need to be conducted more thoroughly than for a target associated with permissive licensed projects. The co-founder and head of corporate development explained how he spends a lot more time on evaluating software associated with copyleft licensed OSS than software associated with permissive licensed OSS when evaluating a potential target: *"Yeah, so I put a lot of attention on the copyleft. To make sure that it's being used right. For permissive? It's like, don't even worry about it. So yeah, a huge amount of attention goes into the use of copyleft."* The CEO of an acquirer explained why acquiring targets associated with copyleft licensed projects requires a lot more planning: *"Anything GPL touches gets infected by it. And so that means your own engineering teams working on that project has to be very, very aware and very careful of what they pull in, what they integrate with, what they cross over in the product line. And so typically, I would say a GPL acquisition requires lot more integration planning than an ASL v2 [Apache License Version 2.0] acquisition. [...] It does require a super amount of care so that you could say that that enforces planning, good planning.*" A chief architect, who has been part of OSS-related acquisitions as a manager at an acquirer and a target, explained how this is particularly relevant for targets with a large codebase: "*The Open Source can be kind of a wild west in that regard [licenses]. And so, it's not your direct dependencies, it's the dependencies of the dependencies and in big codebases that have lots of sprawling dependencies, it can be a super challenge to kind of get through all of that mud to really kind of tease out the licensing.*"

Result 6. Acquirers who want to integrate the acquired software with their own software seem to prefer targets associated with permissive licensed projects over copyleft licensed projects. Furthermore, acquirers spend more time on evaluating software associated with copyleft licensed OSS than software associated with permissive licensed OSS.

3.4 Discussion and Conclusion

3.4.1 Summary of Findings

The goal of this qualitative study was to examine how the fact that a firm is active in OSS influences the selection and evaluation of this firm as an acquisition target. Concretely, I first aimed to explore why firms active in OSS get acquired and which OSS-specific acquisition motives exist. Second, I aimed to understand if and how the engagement of the target in OSS, where development processes, including the feedback from the contributor and user community are well observable, can influence target evaluation processes. To do so, I conducted 52 interviews with acquirers, target employees, and community members across acquisitions in different organizational contexts and countries. I then coded and analyzed the interviews and associated secondary information sources linked to the respective acquisitions. I identified three relevant clusters of codes: the role of OSS in acquisition motives, the role of OSS in the process of target selection and evaluation, and the role of OSS licenses for such acquisitions.

Central to this research project, I found that OSS-related acquisition motives exist. In fact, a large share of the firms was acquired for such reasons. I was able to identify six different OSS-related motives and described how those motives are more or less prevalent depending on the level of OSS activity of the target and the acquirer. It became clear that OSS-related acquisition motives are more prominent in case of targets with a high level of OSS activity, targets that follow an OSS-based business model, and/or targets that have successfully built a large community around them than in case of targets with lower levels of OSS activities and not following an OSS-focused business model. Particularly, the motive of acquiring access to a specific OSS community was only relevant for targets with a high level of OSS activity that had central positions in a community and a community that was actively contributing to the project of the target. Acquirers active in OSS rather focus on acquiring specific OSS talent and leading positions in specific communities. In contrast, non-OSS-active acquirers are less focused on specific OSS communities and rather buy access to OSS-related talent and OSS culture to "learn OSS" from the acquired target or benefit from the brand the target has built in OSS.

From an acquisition process perspective, I found that OSS can be used as a channel to source acquisition targets and as a source of information for target evaluation. Both require the acquirer to be knowledgeable about OSS development itself. Regarding target evaluation, I was able to show that obtaining different information about the target's OSS activities can be valuable for acquirers to reduce technology, market, and integration risk associated with acquisitions.

Lastly, I found that OSS licenses employed by a potential target are an important topic for acquirers in target selection and evaluation. From a strategic perspective, permissive licensed software can more easily be integrated into an acquirer's own software. Also, differences exist in the potential to commercialize the target's software depending on the licenses employed. From a process perspective, interviewees highlighted that technical due diligence and planning require more effort for copyleft licensed projects than for permissive licensed projects.

3.4.2 Contribution

Besides being the first study to show different aspects of the relevance of OSS development for acquisitions, I believe this study contributes to several other research streams.

Acquisition motives. The study contributes to research on acquisition motives (Chatterjee, 1986; Cunningham et al., 2020; Grimpe and Hussinger, 2008; Hitt, Hoskisson, Johnson, and Moesel, 1996; Ouimet and Zarutskie, 2012; Seth, 1990; Walter and Barney, 1990; Worek et al., 2018) by extending the list of known motives with additional motives related to OSS activities. Concerning the known motives of acquiring a firm to gain access to its resources, this study shows that the external community can be of value not only for the firm engaging in the community (Dahlander and Wallin, 2006) but also for an acquirer that chooses to acquire the firm to get access to this external resource. Furthermore, the study adds to the research on motives related to pre-empting competition. Previous research covered pre-empting competition to increase entry barriers (Grimpe and Hussinger, 2008) or "kill" emerging competitors (Cunningham et al., 2020). This study adds to this research showing that—though this rarely happens—firms central to OSS projects may get acquired to exclude competition from using the free OSS, which is a new variation of the pre-emption of competition motive.

Information asymmetry between target and acquirer. This study adds to the research on information asymmetry between target and acquirer (Capron and Shen, 2007; Hussinger, 2010; Shen and Reuer, 2005) by showing that OSS activity of the target can reduce information asymmetry between target and acquirer as the quality of code and talent and the interaction of the target with the community are better observable than in proprietary software development. While prior research has shown that technology acquisitions are characterized by significant information asymmetries between targets and acquirers (Hussinger, 2010) and that those are particularly strong in acquisitions of small companies (Capron and Shen, 2007; Shen and Reuer, 2005), this study suggests that OSS activities of the target, contingent on the acquirer's ability to interpret those, can help to reduce technology, market, and integration risk in technology acquisitions. While a target's proprietary software only allows the acquirer to imprecisely observe the functionality of the software and does not allow the acquirer to observe the quality of the code, OSS allows the acquirer not only to see the contribution of each employee to the code and whether that code "worked," but also if the quality standards of software development have been adhered to, whether the task was challenging, if problems were addressed in a clever way, and so forth.

Licenses in markets for technology. Finally, this study adds new aspects to research on acquisitions in markets for technology (Arora, Fosfuri, and Gambardella, 2001). Specifically, it adds the aspect of the target's choice of licenses for its IP influencing the acquirer's decision to acquire the target and the decision-making process leading to this decision. The interviews showed that targets engaging in permissive licensed OSS projects are preferred because copyleft licenses restrict the uses and integration possibilities of the software. For the same reasons, potential targets engaged in projects under copyleft licenses need to be evaluated more carefully since it is harder to establish if the project can be used in the way envisaged by the acquirer.

3.4.3   Managerial Implications, Limitations, and Outlook

Managerial implications. The findings in this study have several managerial implications. Founders and their investors who consider selling their firm as exit
strategy need to be aware of the potential influence of their firm's OSS activities and their license choice on the acquirer's perception of the firm as a potential
target and how acquirers will evaluate their OSS activities. Particularly, when
initiating an OSS project, the founders need to consider that acquirers seem reluc-
tant to acquire copyleft licensed projects. When the project evolves, founders need
to be aware of the relevance of the code quality and architecture of their OSS projects and the interactions with the community on a technical and personal level
for the evaluation of their firm as a potential target. Vice versa, acquirers need to
be aware of the potential OSS development of potential targets offers for reducing technology, market, and integration risks associated with acquisitions. They
furthermore need to consider that exploiting this potential requires acquirers to
be knowledgeable about OSS development themselves, which typically requires
acquirers to be active in OSS at least to a certain level prior to the acquisition.

Limitations. This study is not without limitations. First, and with regard to the
scope of the acquisitions at hand, the external validity of results is limited. I selected my sample of acquisitions by searching for acquired firms active on GitHub.
This limits the sample both in terms of scope, as other platforms are available and
some firms publish their OSS on their own, without involving a platform such as
GitHub or SourceForge, and years covered, as GitHub data is only available after
2010 and my sample actually only covers acquisitions from 2014–2018. While I
do not foresee many differences between OSS development on GitHub compared
to other platforms, the years covered should be taken into account when transferring the given findings to other acquisitions in earlier years. In the early years
of OSS development, many firms were much more sceptical about OSS development, which might imply a different view of acquirers towards potential targets
active in OSS development. Second, and with regard to the selection of interview
partners, internal validity is limited. While I cover a large number of acquisitions, the number of interviews per acquisition is limited, and, in some cases, only
a single informant was interviewed per acquisition. Furthermore, comparatively
few managers of acquirers were interviewed compared to employees of acquisition targets. Thus, a larger number of executives from the acquirer side involved
in acquisitions would diversify and solidify the results. Third, this study captured
why firms active in OSS get acquired, but it does not capture all the cases where
a firm was not acquired because of its OSS involvement.

Future research. This study represents the first evidence about the role of OSS development in acquisitions. It elucidates central questions about the role of OSS in target sourcing, evaluation, and selection. However, additional research is required to also understand the drawback of a potential target's involvement in OSS in the context of acquisitions. As mentioned in the limitations, this study does not cover cases where an acquirer decided against an acquisition because of the involvement of the potential target in OSS development. Furthermore, particularly quantitative evidence would help solidify these initial results and better understand how different aspects of a target's OSS activities, such as its license choice and interaction with the community, influence target selection. Another interesting factor would be to understand the role of OSS development in the timing of technology acquisitions (Fischer et al., 2020; Hlavka, 2019; Stein, 2017). The present study suggests that OSS activities of the target can reduce risks associated with acquisitions, a central factor in the timing of acquisitions. The timing of acquisitions where targets are active in OSS might differ from acquisitions where targets focus on proprietary technologies. Lastly, the study at hand focuses on the role of OSS before acquisitions. Further research is required to understand the consequences of acquisitions on OSS development, particularly how OSS development of targets, acquirers, and the community around them evolves after acquisitions.

Quantitative Study: The Role of OSS for Likelihood and Timing of Acquisitions

4

4.1 Introduction and Motivation

The previous chapter's qualitative study suggested that firms' OSS activities can be relevant for acquisition decisions. The study presented in this chapter attempts to generate first *quantitative evidence* for whether participating in open innovation ecosystems, such as OSS development, plays a role in acquisitions. Specifically, the study aims to uncover differences between firms active in OSS development and other firms active in software development but not active in OSS, as well as differences within the group of firms active in OSS development. It also aims to generate initial evidence about the role of the acquirer's engagement in OSS prior to an acquisition.

A key decision in acquisitions is which target to acquire and when (Chakrabarti and Mitchell, 2013; Fischer et al., 2020; Kavusan et al., 2020; Ransbotham and Mitra, 2010; Rogan and Sorenson, 2014). My interviews in Chapter 3 suggest that firms' OSS activities can influence their attractiveness as potential acquisition targets. Yet, it is not clear whether participating in OSS makes a target attractive—and, therefore, more likely to get acquired. On the one hand, being active in OSS can have several advantages for a firm when it comes to its (potential) acquisition. For example, being active in OSS may allow firms to gain competitive advantages over firms not active in OSS, such as faster adoption of their product (West, 2003), attracting more valuable talent (Ågerfalk and Fitzgerald, 2008; Henkel, 2004), or cutting product development costs (Dahlander and Magnusson, 2005; Gambardella and von Hippel, 2019), a factor particularly relevant for young firms, which

---

Electronic supplementary material The online version of this chapter
(https://doi.org/10.1007/978-3-658-35084-0_4) contains supplementary material, which is available to authorized users.

© The Author(s), under exclusive license to Springer Fachmedien Wiesbaden GmbH, part of Springer Nature 2021
M. Vetter, *Acquisitions and Open Source Software Development*, Innovation und Entrepreneurship, https://doi.org/10.1007/978-3-658-35084-0_4

often lack access to a large pool of resources (Gruber and Henkel, 2006; Hepp, 2016). These advantages, in turn, can make the firms more attractive targets. Furthermore, my interviews suggest that acquirers can learn about a potential target by tracking the target's activities and their community in the OSS ecosystem. This knowledge might be useful for acquisitions, where gaining sufficient knowledge about a potential target is a key challenge for acquirers and a common reason for many acquisitions to fail. Learning about a target can reduce uncertainty about a target's resources and capabilities (Hlavka, 2019; Ransbotham and Mitra, 2010), and, therefore, ceteris paribus make targets active in OSS more attractive.[1] On the other hand, participating in OSS development might also send less favorable signals to acquirers: for example, many young firms mainly engage in OSS development due to a lack of resources (Gruber and Henkel, 2006; Hepp, 2016) or struggle to monetize their OSS (Chesbrough and Appleyard, 2007; Dahlander and Magnusson, 2008; West and Gallagher, 2006). Also, being active in OSS is associated with risks not present in proprietary software development. First, competitors can easily work with ideas contained in a firm's OSS at little to no cost, which can result in a loss of competitive advantage (Henkel, 2006; West and Gallagher, 2006). Second, a firm that originated the OSS might even lose control over its further development if the community decides to continue a fork of the project and excludes the firm from participating (Raymond, 2001). All these downsides need to be considered by a potential acquirer and may lead them to be cautious about acquiring targets active in OSS development. Hence, OSS literature does not allow us to make clear predictions on how acquirers value a potential target's OSS activities. Neither does acquisition literature, which has so far not touched upon the openness of targets as a factor influencing acquisition decision-making.

In this absence of clear theoretical predictions, I depart from traditional hypothesis-testing, focusing instead on rigorous quantitative analysis in an exploratory fashion (Oxley, Rivkin, and Ryall, 2010). I explore the role of OSS engagement of firms for acquisitions in three steps. First, I explore differences between firms active in OSS and firms not active in OSS and their probability of getting acquired using a survival model. Second, I explore the role of different characteristics of a firm's OSS-engagement, namely the firm's overall commit activity, a firm's ability to attract contributions from the community, and the licenses they focus on, and their association with changes in a firm's probability of getting acquired. To do so, I again use different survival models. In the last step,

---

[1] See Stein (2017) and Hlavka (2019) for an explanation how reducing uncertainty can lead to earlier acquisitions and Ransbotham and Mitra (2010) who show that patenting activity is a positive signal towards potential acquirers leading to a higher probability of getting acquired.

I explore the role of an acquirer's engagement in OSS prior to an acquisition for the timing of acquisitions using ordinary least squares models. For these analyses, I create a dataset combining firm and acquisition data from Crunchbase with OSS development data from GitHub. Focusing on young firms active in software development, I identify acquisition targets' OSS development activities on GitHub. In total, out of 67,793,894 firms founded between 2011 and 2018, I identified 12,597 firms active in OSS; In total, 2,639 of the firms in the sample got acquired; 808 of those were active in OSS development when they got acquired. For the last step of my analysis, I additionally collect OSS development data and other data about the acquirers of the acquisition targets.

I find that firms being active in OSS development are associated a higher acquisition hazard than firms not active in OSS. Within the group of firms active in OSS development, I find that higher levels of OSS activity are associated with a higher acquisition hazard. When controlling for potentially confounding variables, I do not find evidence that license choice or a firm's ability to attract outside contributions significantly influences its probability of getting acquired. Lastly, I find that OSS acquirers acquire targets active in OSS significantly earlier than acquirers not active in OSS. I conduct a series of robustness checks to ensure the validity of the results.

I discuss potential mechanisms behind these findings in the context of existing OSS and acquisition literature and findings from the interviews. My findings make several contributions to OSS and acquisition research and thereby adds to a call to produce novel theory at the intersection of open innovation and strategy research (Alexy, Frederiksen, and Hutter, 2020). Specifically, I contribute to research on target characteristics influencing target selection and acquisition likelihood (Chakrabarti and Mitchell, 2013; Fischer et al., 2020; Hernandez and Shaver, 2019; Ransbotham and Mitra, 2010; Rogan and Sorenson, 2014), showing that firms active in OSS are associated with a higher likelihood of getting acquired. The study also adds to research on the role of acquirers' capabilities for the timing of acquisitions (Hlavka, 2019), showing that acquirers active in OSS themselves can get earlier access to targets active in OSS and build specific capabilities, making it easier to evaluate targets. The study also adds to research on OSS, and broader open innovation, by showing how a firm's openness can influence strategic acquisition decision-making and by expanding the list of capabilities firms can build from being active in OSS development (Dahlander and Wallin, 2006; Nagle, 2018a)

4.2   Theoretical Background

To lay the foundations for this study, I review the literature on factors influencing decision-making in the process towards acquisition decisions focusing on factors influencing target selection, acquisition likelihood, and acquisition timing. Additionally, highlighting key literature from OSS research, I will then show how current literature cannot make clear predictions about the role of a young firm's OSS activities for acquisition decisions.

4.2.1   Factors Influencing Target Selection, Acquisition Likelihood, and Acquisition Timing

Which firms actually get acquired is a key question in acquisition research. Choosing an appropriate target is key for the success of acquisitions (Haspeslagh and Jemison, 1991), but it is not easy. Potential targets may vary in their quality as well as their complementarity with the acquirer. Many of the aspects on which acquirers would like to evaluate potential targets are difficult to assess even after thorough due diligence as acquirers face an information asymmetry (Chondrakis, 2016; Hansen, 1987; Rogan and Sorenson, 2014). Founders, owners, and managers of a potential target understand their firm's strengths and weaknesses better than the acquirer (Rogan and Sorenson, 2014). They might also understand the market environment they operate in better than an acquirer not yet active in this product or regional market. Furthermore, those founders, owners, and managers regularly have their own goals, and might even wish to promote poor acquisitions before evidence of the flaws of the target becomes apparent to the acquirer (Rogan and Sorenson, 2014). As a result, acquirers value information that helps to decrease uncertainty about a potential target's quality. Several researchers have found that signals indicating higher quality of a potential target increase acquisition likelihood. For example, signaling higher technical quality through patents was found to significantly increase acquisition likelihood (Ransbotham and Mitra, 2010). Similarly, Fischer et al. (2020) show that acquisition likelihood strongly increases when technological uncertainty declines as a result of regulatory approval of a new product in the medical device industry. Research also found that—given the presence of public and private alternatives—acquirers tend to choose public targets over private targets, particularly if the industry is new for the acquirer, the targets have significant intangible assets, or if the targets have not signaled their value for an acquisition through collaborations or in other ways (Capron and Shen, 2007; Shen and Reuer, 2005). In line with this argument, direct ties between a potential target and its acquirer or common ties between them, e.g., in the form of common clients, can also help to overcome information asymmetries and increase the probability of the potential target to get acquired (Hernandez and Shaver, 2019; Rogan and Sorenson, 2014; Zaheer et al., 2010). Lastly, collecting information on targets becomes more difficult with an increasing geographic distance between the target and the acquirer, resulting in a lower probability of an acquisition to take place (Chakrabarti and Mitchell, 2013).

The role of uncertainty around a potential target, its technology, and the market it operates in has also been highlighted in the literature on acquisition timing. Research found that information asymmetry around a target's qualities is particularly present for younger targets, which are in the focus of this study (Hussinger, 2010). Acquirers face a trade-off between uncertainty and higher acquisition prices when deciding if they want to acquire a target earlier or later (Sorenson and Stuart, 2008). The earlier in its lifecycle a target is bought, the less expensive it usually is. However, the younger the target, the less clear it is whether the target's product or technology will eventually succeed, and if there is a market for the product or technology. As a result, younger targets are associated with more uncertainty than older ones (Shen and Reuer, 2005). Hlavka (2019) showed that the acquirer's capabilities to evaluate targets, such as experience from previous acquisitions, can help the acquirer reduce uncertainty in acquisition decision-making and lead to younger targets (Hlavka, 2019). Furthermore, increased demand for targets within a certain field due to a technology hype makes acquirers accept higher levels of uncertainty (Hlava, 2019). Acquirers also tend to buy targets possessing certain technologies the acquirer currently lacks earlier if they expect this technology to become part of a standard (Warner et al., 2006). Additionally, research has also shown that in the market for novel technologies, followers get acquired earlier relative to their founding date than first movers, as they can benefit from the progress the first mover has made in reducing the uncertainty surrounding a novel technology (Fischer et al., 2020).

Next to signals reducing uncertainty, additional factors influencing acquisition likelihood and timing have been identified by prior research. Kavusan et al. (2020) show that acquisition likelihood increases with increasing technological similarity and complementarity based on patents of target and acquirer. They argue that the similarity of an acquirer's and a target's technology makes it easier for an acquirer to evaluate a target, while the complementarity of their technology gives the acquirer potential to develop novel innovations. Yu et al. (2016) further differentiate between similarity and complementarity of R&D pipelines and product portfolios and show that acquirers prefer similar targets when it comes to R&D pipelines, while they prefer complementary targets when it comes to product portfolios.

From a capability-based perspective, acquirers strongly prefer targets inferior to themselves in existing geographic markets but are more likely to choose superior targets in new markets (Kaul and Wu, 2015). Additional factors that have been shown to positively influence acquisition likelihood are a favorable position of the potential target in a network of firms (Hernandez and Shaver, 2019) and similarity in national culture between the acquirer and the potential target (Rao et al., 2016). With a focus on timing, Stein (2017) has shown that acquisition targets are younger and smaller in technology acquisitions focusing on performance improvement of the acquirer's technology portfolio compared to acquisitions focusing on adding additional functionalities to the acquirer's technology portfolio.

In summary, prior research on acquisitions has addressed which characteristics of the target, the acquirer, and the market influence decision-making in acquisitions from several perspectives. It has particularly highlighted the role of uncertainty for such decisions and how factors reducing uncertainty can help acquirers in their decision-making processes. However, it has not addressed aspects of the role of a target's or acquirer's engagement in OSS development, or generally open innovation, for acquisition decision-making.

4.2.2   Firms' Potential Benefits and Downsides of Engaging in OSS

Similar to the acquisition literature, open innovation and OSS literature has not yet covered the intersection of OSS and acquisitions. However, it can still be informative for research at this intersection. Particularly, I want to cover literature that has elaborated on the potential benefits and downsides of engaging in OSS development for firms, as these benefits and downsides should also be key factors for an acquirer to consider when acquiring a firm.[2]

Potential benefits of engaging in OSS. Firms engage in OSS development and open source their code for a variety of reasons, such as access to feedback, novel ideas and technologies, access to talent, to develop new capabilities, or a faster diffusion of their technology (Bonaccorsi et al., 2006; Dahlander and Magnusson 2005; Raymond, 2001; Spaeth et al., 2015). Research has shown that firms can develop strategic advantages from their engagement in OSS. For example, OSS can enable firms to gain faster user adoption (West, 2003), attract valuable talent (Ågerfalk and Fitzgerald, 2008; Henkel, 2004; Lerner and Tirole, 2002), or

---

[2] See Section 2.1.4 for a more detailed description of the general benefits and downsides of a firm's engagement in OSS without considering the context of acquisitions.

increase demand for complementary services and products (Andersen-Gott et al., 2012; Behlendorf, 1999; Hecker, 1999; Raymond, 2001). Being active in OSS can also help them to cut product development or sourcing costs (Dahlander and Magnusson, 2005; Gambardella and von Hippel, 2019), a factor particularly relevant for young firms and younger firms, which often lack access to a large pool of resources (Gruber and Henkel, 2006; Hepp, 2016). Furthermore, Nagle (2018a) found that firms active in nonpecuniary (free) OSS are associated with significant positive value-added returns.

Potential downsides of engaging in OSS. Yet, engaging in OSS does not only have advantages. Contributing to OSS projects always also means giving up some IP, which potentially could have been sold otherwise. As a result, monetizing software that has been open sourced is difficult and requires creative business models to profit from OSS (Chesbrough and Appleyard, 2007; Dahlander and Magnusson, 2008; Perr et al., 2010; West and Gallagher, 2006). Also, giving up IP can potentially mean a loss of competitive advantage as competitors can easily work with ideas contained in a firm's OSS at little to no cost (Henkel, 2006; West and Gallagher, 2006). Furthermore, building a community and motivating the community to supply an ongoing stream of external innovations and feedback is a key challenge for firms active in OSS (Daniel et al., 2018; West and Gallagher, 2006). Building a community requires time, and firms must walk a fine line when desiring to obtain financial benefits from their investments in the OSS (Schaarschmidt, Walsh, and von Kortzfleisch, 2015), as the OSS community will less likely engage in the development of the software if the firm is perceived as being driven primarily by profit motives (Stewart et al., 2006). A firm that originated the software might even lose control over its further development if the community decides to continue a fork of the project and excludes the firm from participating (Raymond, 2001). An acquirer faces the same risk after an acquisition: the community could fork an acquired project and continue its development without the new owner of the acquired target if they disagree with the acquisition or the new owner. All these risks need to be considered by a potential acquirer. But even if the firm is not following a strategy where major parts of its software are available "in the open" and for free, just relying on OSS for their own (proprietary) software has potential drawbacks when it comes to acquisitions. Particularly utilizing copyleft licensed OSS, which requires derived work to be copyleft licensed OSS as well, is relevant here. Research has shown that firms do not always correctly label OSS they utilized for their software development (Madanmohan and De', 2004). For example, an industry study by Black Duck Software, one of the largest providers for OSS compliance management and OSS due diligence, reported that 47% of companies do not have formal processes in place to track OSS code utilized in their codebase (Black Duck Software, 2016). As a result, acquirers might end up acquiring software that is actually infringing on an OSS license. This may reduce the value of the acquired IP of the target, which was otherwise considered to be licensable under a proprietary license, significantly.[3] Lastly, when it comes to an acquisition of a firm, its engagement in OSS might turn out as an unfavorable signal of resource scarcity: Engaging in open innovation is known to be a method to overcome resource scarcity (Dahlander and Magnusson 2008; Franke and Shah 2003; Gruber and Henkel 2006; Harhoff et al., 2003; Henkel, 2006; Hertel et al., 2003; Lakhani and Wolf, 2005), which is particularly relevant for small and young firms (Hepp, 2016). Hence, acquirers might interpret the activities of firms in OSS as a signal of not having been able to build the respective resources internally.

Summarizing, their engagement in OSS can make firms attractive targets if they are successful in leveraging those potential advantages of engaging in OSS. Yet, participating in OSS also comes with significant risks and potential downsides, and not all firms are successful with their engagement in OSS (Capra, Francalanci, Merlo, and Rossi-Lamastra, 2011; Daniel et al., 2018; Stewart and Gosain, 2006). Hence, OSS literature does not allow us to make clear predictions on how acquirers value a potential target's OSS activities. They might, for example, consider OSS a promising and novel tool to increase user adoption or utilize a target's OSS as a way to decrease uncertainty around a target's technological capabilities and its adoption by users (as pointed out in the interviews in Chapter 3), which might result in more and earlier acquisitions of firms active in OSS development. But they might also see a potential target's OSS activities as a risky business with a potential lack of opportunities to monetize the target's innovations or see the community "run away" after the acquisition altogether, which would decrease the attractiveness of the firm as a target. Making clear predictions on how acquirers value a target's OSS activities is even harder in the presence of novel potential motives to acquire firms active in OSS development not present in firms focusing on proprietary innovation, such as acquiring a firm with the intention to shut down their OSS project to eliminate a competitive threat originating from the open, and often free software.

Taken together, existing literature on acquisitions has uncovered several mechanisms that influence target selection, acquisition likelihood, and acquisition

---

[3] The practical relevance of this topic is highlighted, e.g., by a checklist for OSS license due diligence in acquisitions released by the Linux Foundation (https://www.linuxfoundation.org/open-source-management/2019/03/assessment-open-source-practices/; last accessed on 02.12.2020) or commercial offerings for OSS license due diligence, e.g., by Synopsis (https://www.synopsys.com/software-integrity/open-source-software-audit.html; last accessed on 02.12.2020).

timing. However, it has not covered the role of openness of targets and acquirers in acquisition decision-making. OSS literature has highlighted not only many potential benefits that make firms active in OSS attractive targets but also many risks and downsides that are associated with a firm's engagement in OSS. As a result, the literature does not allow us to conclude how firms' OSS activities influence the firms' overall attractiveness and as a result, how likely they are to be acquired. The interviews, as well as prior literature, show that this engagement in OSS could influence such decisions. Therefore, I depart from traditional hypothesis-testing and instead focus on exploratory, rigorous analysis of data to explore the role of a firm's OSS activities for acquisition-decision making, an approach suggested by Oxley et al. (2010) and currently gaining increasing importance in management literature (e.g., Agarwal, Braguinsky, and Ohyama, 2019).

4.3 Data and Method

In this section, I first explain the selection of data sources. Following, I examine the identification of relevant firms and acquisitions, including how I identify their OSS activities. Lastly, I explain the variables utilized in the analytical models.

4.3.1 Main Data Sources: Crunchbase and GitHub

To investigate the role of OSS development prior to acquisitions empirically, I utilize OSS development data collected from GitHub, which I combine with acquisition and company data collected from Crunchbase.

Crunchbase. For company and acquisition data, I utilize the Crunchbase database. Crunchbase is a curated crowdsourced company database (Dalle, den Besten, and Menon, 2017), that covers company information, such as company size, location, founding date, acquisitions, and the industries a company is active in. The database is widely used in the industry (Hlavka, 2019) and has recently also become more prominent among researchers (Alexy, Block, Sandner, and Ter Wal, 2012; Block, Fisch, Hahn, and Sandner, 2015; Dalle et al., 2017; Homburg, Hahn, Bornemann, and Sandner, 2014; Ter Wal, Alexy, Block, and Sandner, 2016). It is particularly rich in information about young firms and tech companies, which are naturally in the focus of this study given the focus on (Open Source) software. For such firms, the acquisition database of Crunchbase is considered more complete than traditional data sources, which often only focus on the largest acquisitions or acquisitions of public companies (Hlavka, 2019). When

I collected the data in June 2019, the database contained 715,410 companies and 91,426 acquisitions in total.

GitHub. For measuring OSS activities, I selected GitHub as a data source. GitHub is popular among the members of the OSS community, students, and hobbyists (Burton et al., 2017), allowing developers to host and publish their OSS projects and contribute to the OSS projects of other users. Every user on GitHub can download, use, and modify source code in public OSS projects and offer the modifications back to the projects if wanted. It is also used by many software companies, including Google, Microsoft, Red Hat, Facebook, Uber, Intel, IBM, and Alibaba,[4] and many prominent OSS projects such as Node.js and Ruby on Rails are developed on the GitHub platform (Sims and Woodard, 2020). As of this writing, GitHub had 31 million users and hosted over 100 million software projects,[5] and is the largest OSS development platform in the world (Gousios et al., 2014). Whereas a large number of those OSS projects do not attract contributions from outside contributors, other projects are supported by thousands of developers voluntarily contributing to the projects, many of those contributing on a regular basis (Sims and Woodard, 2020). Given the richness and open nature of the data available, GitHub provides a rich empirical setting for management research, allowing observability of the development of projects and OSS work of individuals and companies on a very granular level over time. As a result, GitHub as a data source has recently sparked increased interest among researchers in the field of management (e.g., He et al., 2020; Jiang, Tan, Sia, and Wei, 2019; Nagle, 2019; Sims and Woodard, 2020). Particularly the considerable prominence of GitHub among firms makes this data source well suited for research on firms and their engagement in OSS development. I selected GitHub over other OSS development platforms (e.g., SourceForge) because of GitHub's large number of projects and registered contributors, the ability to track each individual contributor's activities over time (Jiang et al., 2019), the homogeneity of the data (Gousios and Spinellis, 2017), and the possibility to match contributor's email addresses with company domain information, which I obtain from Crunchbase. To create a rich set of information about the OSS activities on GitHub, I use data from three different data sources containing GitHub data. Figure 4.1 gives an overview of the data source for GitHub data used for this research.

---

[4] https://www.freecodecamp.org/news/the-top-contributors-to-github-2017-be98ab854e87/ last accessed 07.12.2019.

[5] https://venturebeat.com/2018/11/08/github-passes-100-million-repositories/ last accessed 12.06.2020.

The primary source is the GHTorrent project (http://ghtorrent.org/).[6] The GHTorrent project is a scalable, queryable, offline mirror of data publicly available on GitHub (Gousios, 2013). Among others, it contains information about each public code commit, user, and project since GitHub's founding in 2008. It is a frequently used data source in computer science research, with more than 400 research articles on Google Scholar citing the data source.[7] I complement this data with data from the GHArchive (https://www.gharchive.org/) and data I obtained directly from the GitHub API (https://developer.github.com/). GHArchive data is used to obtain email-domains used by GitHub users when committing code on GitHub, while I utilize the GitHub API to obtain license information about the projects (i.e., repositories) hosted on GitHub. All three data sources are well known in computer science research (Cosentino, Luis, and Cabot, 2016). The datasets can be matched, as each commit on GitHub has a unique identifier ("sha") included in GHTorrent and GHArchive, and each project has a unique URL, which can be found on GHTorrent and in the GitHub API. I collected data on GitHub activity for the years 2011–2018 (the GitHub Archive is available starting 2011). The acquisition data from Crunchbase was selected to match the timeframe where GitHub data was available.

GitHub Torrent	GitHub Archive	GitHub API
• http://ghtorrent.org/	• https://www.gharchive.org/	• https://developer.github.com/v3/
• Main data source for project	• Contains metadata of all events on public GitHub repositories (e.g. PushEvents (which contain commits), all events of PullRequests, Comments, etc.*)	• Gives access to all publicly available GitHub data
• Project mirroring most metadata of GitHub and saves it in a structured manner that allows tracking most activities of a user or in a project over time (particularly commits)		• Rate limit of 5.000 requests/h limits usability of data source
	• Used in this project to collect email domains used by the creator of a commit (not contained in GHTorrent)	• Used in this project to collect project licenses and for manual check of validity of approach and other data sources
• Allows retrieving information of deleted repositories (e.g. license information)	• Commit domains can be matched to GHTorrent via commit-"sha"	• Matched to other data sources via unique project or user URLs
• Data stored since start of GitHub in 2008 and updated monthly ending mid 2019 when the project paused until early 2020	• Data stored since 2011, updated hourly	• Updated real-time, but deleted files cannot be viewed

*For event types see: https://developer.github.com/v3/activity/events/types/

Figure 4.1 Data sources utilized for GitHub data

---

[6] I am grateful to Georgios Gousios, founder of the GHTorrent project, for giving me the opportunity to discuss my approach with him and granting me access to his GHTorrent MongoDB server, giving me access to over 12TB of raw GitHub data without being required to replicate the full database ourselves (available for download here: http://ghtorrent-downloads. ewi.tudelft.nl/mongo-daily/ as of 12.06.2020).

[7] https://scholar.google.gr/scholar?cites=11132126230347149781; last accessed 12.06.2020.

4.3.2 Sampling of Firms and Identifying Their GitHub Activity

The main analysis presented in this chapter relies on a survival analysis estimating the acquisition hazard of firms depending on their OSS activities and several controls. In this section, I describe which filters I applied for the sampling of firms for the analysis and how I identified their GitHub activity.

As GitHub data is limited to the years from 01.01.2011 onwards, the focus of this study naturally lies on young firms founded in 2011 and later. Firms founded after 31.12.2018 or those without a record for their founding date are also excluded, as this is where my GitHub data ends. Focusing on young firms has the advantage that one can assume that participating in OSS is a major strategic decision for the firm, and, hence, an important point to consider for the acquirer as well. Young firms have very limited (human) resources, and their activities are more focused on one or a few products; hence, if the firm decides to get active in OSS development, one can consider this to be a major and deliberately made decision by the firm, one that has most likely been made by the leadership of the firm (as opposed to a large corporation where a department decides to participate in OSS without the top-management's awareness). Focusing on young firms is also of practical significance, as young firms are critical to economic growth and acquisition of such firms is a major component of the strategy of many companies (Graebner and Eisenhardt, 2004).

To further ensure that being active in OSS is actually a relevant option for a firm and a key decision for its leadership, I focus on firms active in software development. As acquisitions are an integral part of the growth and innovation strategy, particularly in the software development sector, focusing on young firms from this sector is especially suitable for research on acquisitions. To select firms active in software development, I utilize Crunchbase's company industry category tags assigned to each company and filter for firms assigned to the Crunchbase "software" industry group.[8] The number of category tags per company firm ranges from one to ten. A firm is included in the sample if one of the category tags is within the "software" industry group. According to Hlavka (2019), the category tags are a unique advantage of the Crunchbase dataset, as they go into much greater detail than, e.g., the commonly used Standard Industrial Classification (SIC) codes (e.g., used in Carow, Heron, and Saxton, 2004; Hayward, 2002; Zollo and Singh, 2004). However, for a robustness check, I use SIC codes as

---

[8] The Crunchbase industry group "software" contains 90 software related software-related sub-industries such as "Enterprise Software", "Online Games", "SaaS", "Developer Platform", etc. See https://support.crunchbase.com/hc/en-us/articles/360043146954-What-Ind ustries-are-included-in-Crunchbase- for the full list (last accessed 12.06.2020).

an alternative selection of "software firms"; I focus on firms that are operating under the three-digit SIC code 737 ("Computer Programming, Data Processing, and other Computer Related Services"), which contains the majority of (US) software firms. I collect SIC codes from Orbis and match the data with Crunchbase based on company websites.

Next, I remove all firms from the sample that do not have a record for their number of employees in Crunchbase. I do so as one of the control variables I utilize in my analysis is the size of a firm based on its number of employees.

To identify firms active in OSS, I match email-domains of firms, which I obtain from Crunchbase, with the email addresses GitHub users used on the platform when creating commits. Identifying employees of firms based on the user's email address is an approach used before in management research, for example, by Nagle (2018a). To ensure the relevance of the work on GitHub and avoid duplicates, I only consider work done on organization-hosted repositories (thereby excluding privately hosted repositories), and similar to He et al. (2020), I restrict the sample to independent projects (i.e., not forked from a parent project).[9] As I can only match firms with a domain in their Crunchbase data with GitHub, I exclude all firms without a domain in Crunchbase from the analysis. Furthermore, several domains needed to be excluded manually, as they did not allow the clear identification of their employees. The reasons for not being able to identify them clearly are as follows: I could not include email-providers, such as yahoo (acquired by Verizon in 2017), in our dataset, as most commits under the "yahoo.com"-domain are from users of the yahoo email service, not employees. For a similar reason, I excluded online education facilities, such as the Flatiron School (acquired by WeWork), as they often use GitHub as a host for their students' projects, as well as companies using a LinkedIn or Facebook page as their homepage or a government or a university domain on Crunchbase. I also excluded firms that used their parent organization's domain (prior owner in case of a spinout or carve-out or the acquirer) on Crunchbase, as I cannot differentiate between those employees affected by an acquisition and those not part of the acquired entity.

For the main survival analysis, I focus on firms that remained active until 31.12.2018 or got acquired. Hence, I exclude firms that stopped operating or went public. To avoid a potential selection bias, I also run a competing risk analysis with IPOs and "stopped operating" as alternative events to acquisitions as a robustness check (see Section 4.4.2). For this robustness check, I needed to drop

---

[9] I exclude forks since they inherit all contributions from their respective parent project and hence create a large number of duplicates.

an additional 751 of the firms which ceased operations, for which no date when they ceased operations could be obtained (see next section for details).

Table 4.1 summarizes the sampling and cleaning steps. Overall, the sample contains 67,794 firms founded between 2011 and 2018 for the analysis. 2,639 (3.9%) of those were acquired between 2011 and 2018, 111 went public (0.2%), and 3,172 ceased operations (4,7%). Overall, 13,036 firms in the sample have been active in OSS during the observed period (19.2%). 808 of those have been active in OSS in the year prior to their acquisition. In total, the panel dataset contains 3,514,648 monthly observations (on average ~ 51.9 observations per firm), capturing the firms' OSS activities on a monthly basis.

Table 4.1 Sampling steps and sample size

Sampling step	Nr. of firms before	Nr. of firms after	Firms excluded in step
(1) Firms founded between 2011 and 2018	715,410	238,878	476,532
(2) Firms in "software" industry	238,878	82,307	156,571
(3) Firms having a homepage	82,307	80,347	1,960
(4) Semi-manual exclusion of domains not allowing employee identification in GitHub	80,347	80,217	130
(5) Firms having a company size record	80,217	68,545	11,672
(6) Exclusion of firms ceasing operations but not having a date when they did so	68,545	**67,794**	751
(7) Exclusion of firms going public or ceasing operations (*optional; depending on the analysis*)	67,794	**65,262**	2,532

Bold: Samples relevant for the analysis

4.3.3 Variables

In this section, I explain the definitions and data sources for my variables, including all necessary dependent variables, independent variables, and controls. The section starts with the variables describing the focal firms in the sample (i.e., the set of potential targets and the firms that got acquired) and ends with variables concerning the acquirers, which are only used in the last step of my analysis. As I utilize different quantitative models to explore the phenomenon at hand, I will explain the models employed in the results section in more detail together with the results of the respective model. Table 4.2 gives an overview of all variables utilized in this chapter.

The main dependent variables I utilize in this chapter are *Event: Acquisition* and *Firm Age* in months. Acquisition event is a binary variable per firm that is 1 in a month where a firm got acquired and 0 otherwise. The firm's age in months is calculated relative to its founding date. The variable increases by 1 for each month since the firm's founding date and ends with an (acquisition) event or at the end of the observation period. For some models, only the firm's age in months at the acquisition event is utilized. The data is available in Crunchbase. For a competing risk model, which is one of the robustness checks, there is an additional nominal variable per firm called *Event* that takes the value 1 in a month where an acquisition event took place, 2 in a month where a firm ceased operations, 3 in a month where a firm went public, and 0 if no event took place. Data for

Table 4.2 Overview of variables used in quantitative analysis

Variable	Unit	Description
Firm Age	Months	The firm's age in months relative to its founding date
Event	Multinomial variable	(0) No event (1) Acquisition takes place (2) Operations ceased (3) IPO of firm
Event: Acquisition	Binary	(0) If no event takes place, (1) if acquisition takes place
FirmOSS	Binary	(1) if employees of the firm were contributing to non-private OSS projects on GitHub, (0) if no activity on GitHub can be associated with firm.
Level of OSS activity	Double	Log-transformed number of contributions (+1) of a firm's employees in the one-year period prior to a (potential) acquisition.
PullRequests	Binary	(1) if the firms received a pull request from a user not employed by the company in the year prior to a potential acquisition, (0) otherwise
License	Binary	(1) if the firm focuses on permissive licensed projects (e.g., MIT, Apache, BSD), (0) if the firm focuses on copyleft licensed projects (e.g., GPL, AGPL, LGPL).

(continued)

Table 4.2 (continued)

Variable	Unit	Description
LnPatents	Integer	Log-transformed aggregate number of patent applications filed until focal month (+1)
HasInvestors	Binary	(1) if the company received investments from an investor, (0) otherwise
EstFirmSize	Double	Firm size by employees in each month. Estimated based on the growth rate in Crunchbase categories per month.
LnRevenue	Integer	Log-transformed revenue in $ from Orbis
AcquirerOSS	Binary	(1) if employees of the firm were contributing to non-private OSS projects on GitHub, (0) if no activity on GitHub can be associated with firm.
Collaboration	Binary	(1) if employees of target and acquirer contributed to the same projects in the year prior to the acquisition, (0) if not.
Acquisition Experience	Integer	The number of acquisitions the acquirer has made in the five years prior to an acquisition
AcquirerSize	Nominal variable	Number of employees of buyer, in nine pseudo-logarithmic size classes as reported by Crunchbase with the following employee ranges: (1) 0 to 10 (2) 11 to 50 (3) 51 to 100 (4) 101 to 250 (5) 251 to 500 (6) 501 to 1,000 (7) 1,001 to 5,000 (8) 5,001 to 10,000 (9) 10,001+
Country Overlap	Binary	(1) if acquirer and target reside in the same country, (0) otherwise
Industry Overlap	Binary	Pairwise comparison of the list of Crunchbase's acquirer industry categories and target industry categories. (1) if at least one category is identical, (0) otherwise

IPOs with the respective IPO date is available in Crunchbase ($n = 111$ IPOs in the sample). Data for the date a firm ceased operations is more difficult to obtain. While Crunchbase records if a firm ceased operations, only 36.5% of the record contain a date when they did so ($n = 3,172$ firms which ceased operations, thereof 1,159 with a date when they did so in Crunchbase). To improve the availability of this data, I utilized Orbis as an additional data source, which yielded closure dates for additional 431 (15.6%) of the firms. Therefore, I created a workaround to approximate the date a firm ceased operations based on the last date the firm was active on Twitter. Twitter proved to be the best social media platform to collect this data, with ~ 65% of the firms that ceased operations but did not contain a date in Crunchbase when they did so being active on Twitter. For a small fraction of the firms, I utilized a firm's Facebook page instead of Twitter or checked their founders' LinkedIn profiles, and approximated when the firm ended operating based on their LinkedIn CVs. Overall, I was able to obtain approximated dates when firms ceased operations for 2,421 firms. The remaining firms for which I could not obtain any information regarding when the firm ceased operations were dropped from the sample ($n = 751$; 24% of all firms ceasing operations).

The main independent variables in the focus of this chapter capture different aspects of the OSS activities of the firms in the sample. In the analysis comparing firms active in OSS with firms not active in OSS, I use a binary variable *Firm-OSS,* which takes the value of 1 if the employees associated with the firm via the domain-based matching explained above[10] (see Section 4.3.2) were contributing to non-private OSS projects on GitHub in the one year prior to a potential acquisition event,[11] and 0 if no activity on GitHub can be associated with the firm. I measure contributions based on commits. A commit is a change in the source code of the project.[12] This measure is an accepted measure for OSS activity, as software development (i.e., changes in the source code) is one of the core community activities (Daniel et al., 2018) and has previously been used in management literature (Daniel et al., 2018; Nagle, 2019). I focus on the one year period prior

---

[10] GitHub users are only associated with a firm for the time they use the firm's domain in their commits. Before they start using the domain and once they stop using the domain, I do not consider them an employee anymore. This way, I can ensure that I only capture activities that have happened when the user was employed by the focal firm.

[11] A firm can be acquired every month until it is actually acquired, going public or ceases operations. Hence, the one year period is always calculated relative to the focal month.

[12] In GitHub a commit is always linked to the person who wrote the code, be a user with direct write access to the project or external contributors who have to submit their proposed commits to a project via a pull request.

to a potential acquisition event, as the OSS activities prior to a potential acquisition should—if even—be relevant for an acquirer in the target search, evaluation, and selection process.

To capture OSS activities of firms in a more granular and diverse manner, I utilize three additional variables. I use the *Level of OSS activity* to capture how active a firm was in OSS development. I use the log-transformed number of contributions (+1) of a firm's employees in the one-year period prior to a (potential) acquisition. For robustness checks, I also utilize a four-category categorization of the firm's OSS activity. The four categories are "no activity" (0 commits), "low activity" (1–16 contributions per year; ~ 33% of the firms active in OSS development in the sample at the time of an event or when they leave the observation period without an event occurring), "medium activity" (17–105 contributions per year; ~ 33% of the firms active in OSS development in the sample), and "high activity" (more than 105 contributions per year; ~ 34% of the firms active in OSS development in the sample).

I use a variable termed *PullRequests*, a binary variable aiming to capture if a firm was able to attract contributions from the OSS community (i.e., from people not employed by the company). The dummy takes the value 1 if the firms received a pull request from a user not employed by the company in the year prior to a potential acquisition, and 0 otherwise (see Section 2.1.5 for a more detailed explanation of pull requests). I use pull requests as the variable for analyzing community contributions to ensure the user contributing to a project is not a maintainer or owner of the project. To approximate which projects are initiated by a firm, I only consider projects where the first contribution to the project was made by an employee of the focal firm (unfortunately, there is no data on founder non-privately hosted projects available). Thereby, I exclude projects that were not initiated by the focal firm but joined by the firm, aiming to measure if the firm was able to attract outside contributions to *their projects*, and not *any project* they joined. For robustness checks, I also utilize the log-transformed number of pull requests (+1) instead of the dummy and, as a second alternative, the number of external contributors.

I use a variable termed *License*, which is a binary variable aiming to capture which type of OSS licenses the focal firm focuses on. I use this variable as the license type affects how the software can be used, particularly, whether it can be integrated into proprietary software (permissive licenses) or if it requires derived work to be of the same OSS license (copyleft licenses), which, in turn, influences a potential acquirer's options to include the code into its proprietary software.

Using the Wikipedia categorization of OSS licenses, [13] I create a dummy that takes the value of 1 if the firm focuses on permissive licensed projects (e.g., MIT, Apache, BSD) and 0 if the firm focuses on copyleft licensed projects (e.g., GPL, AGPL, LGPL). In case a firm contributes to both types of licensed projects, the dummy variable is created based on the license type that received the most contributions by the firm. Most projects' licenses are obtained via the GitHub API, where users can label their projects with a label choosing from the most common OSS licenses. [14] For projects not labeled with such a license-label, I collected additional license data directly from GitHub repositories (readme or license files) and by searching the projects' commit history (in particular for defunct projects via GHTorrent's MongoDB database). [15] For those, I utilize a keyword-matching logic, where I assign licenses based on matches of the text in license files, readme files, or commits to a list of keywords I created based on a list of OSS licenses and their abbreviations published by the Open Source Initiative. [16] In my sample, 69% of firms active in OSS development are associated with permissive licenses and 10% with copyleft licenses at the end of the observation period. [17] For the remaining 21%, I could not find clear license information, either because there was no license information in the projects available or the license type was unclear (e.g., in case of mixed licenses). Those firms were dropped from analyses where I use this variable. The variable can change over time if firms focus on other projects with different license types over time. This happens for 12.8% of the firms active in OSS development. [18]

Next to those independent variables, I use several controls. I use a control *LnPatents,* which captures the number of patents. *LnPatents* reflects the natural logarithm of the count of patents (+1) a firm has filed up to the focal month in

---

[13] https://en.wikipedia.org/wiki/Comparison_of_free_and_open-source_software_licences (last accessed 24.06.2020).

[14] https://help.github.com/en/github/creating-cloning-and-archiving-repositories/licensing-a-repository (last accessed 24.06.2020).

[15] I thank Georgios Gousios, founder of the GHTorrent project, for giving me the opportunity to discuss my approach to utilizing GitHub data with him and granting us access to his GHTorrent MongoDB server.

[16] https://opensource.org/licenses/alphabetical (last accessed 24.06.2020).

[17] The high share of permissive licenses compared to copyleft licenses seems common for OSS development in the last decade and was confirmed by my interviewees. Non-scientific research on GitHub licenses shows similar results (https://www.kaggle.com/mrisdal/safely-analyzing-github-projects-popular-licenses; last accessed 15.12.2020).

[18] I cannot capture changes of licenses of the same project over time as there is no data available for such license changes. Yet, such changes are rare and I do not expect them to influence the results.

time during the observation period (similar to Fischer et al., 2020). I utilize this
variable as research has shown that patents positively influence acquisition likelihood (Hernandez and Shaver, 2019; Ransbotham and Mitra, 2010). In line with
Hernandez and Shaver (2019), I deemed the date of application as the moment
when a patent potentially becomes relevant for the acquirer's target selection, evaluation, and selection processes. Accordingly, I consider all patents regardless if
they were granted or not. I obtain patent data from Google Patents.[19] To match
patent assignee names and Crunchbase company names, I adopted the name cleaning algorithm developed by Arora, Belenzon, and Sheer (2017) to match NBER
patent data with Compustat company data[20] and made some smaller adjustments
to fit Crunchbase company names. Specifically, some names became too generic
after cleaning, so I addressed those by checking if the assignee country and the
Crunchbase company country are the same, and in some cases, manually rever-
ting back to the original company name, or dropping the record if matches were
clearly wrong.[21] I find patents for 5,536 firms in the sample (8.2% of sample).

I use a control *HasInvestors*, which captures if a firm in the sample received
outside investments from business angels, Venture Capital firms (VCs), or other
(investment) companies. I include this variable as an indicator of the quality of
a company's team and technology (Hlavka, 2019; Ransbotham and Mitra, 2010).
The rationale behind including this variable is an investor selects its investees only
after proper due diligence of evaluating both the target team and its product or
technology and the market potential (Donnelly, 2009; Hlavka, 2019; Kaal, 2016).
The variable is a binary variable per firm, taking the value 1 if the firm has received investments and 0 if it did not. Crunchbase lists all investors, yet, in line with
Hlavka (2019), I use a binary variable as the due diligence of a single investor
is already sufficient to control the quality—if the target was not promising, there
would be zero investors. This data is available in Crunchbase. The variable is con-
stant over time as Crunchbase data does not record when an investment was made.
Having more granular data on the timing of the investment would allow a more
granular role investigation of the role of investments, as the investor itself might

---

[19] Publicly available in Google's cloud platform: https://cloud.google.com/blog/products/gcp/
google-patents-public-datasets-connecting-public-paid-and-private-patent-data (last accessed 05.12.2020).

[20] The assignee and company name cleaning logic is available for download here: https://zen
odo.org/record/3594743 (last accessed 05.12.2020). I thank Ali Samei for pointing me to this
method.

[21] E.g., the Crunchbase company name "Element" is contained in more than 1700 assignee
names in Google Patents data, among others "Element Science Inc", "Element Ltd", "Element
Co Ltd", "Element Ltd Corp", "Element KK", "Element Labs Inc", "Element". As a result,
this record was dropped.

take an influence on the decision to sell the firm to an acquirer. Yet, including a binary variable in the analysis is still valuable, as the qualities of the target that made the investor make an investment are already present before the investment is made. In the sample, 20,937 firms received funding (30.9% of the sample).

I use a control *EstFirmSize*, which is used to control for the size of a company. Target size is regularly used as a variable in studies examining acquisition likelihood and timing of acquisitions (Hlavka, 2019; Stein, 2017). Similar to Hlavka (2019), I use Crunchbase data for this variable. As Crunchbase only offers one data point per firm in 9 pseudo-logarithmic categories ranging from "1–10" (equal to a value of 1 on a numeric scale) employees to "10,001 +" (equal to a value of 9 on a numeric scale),[22] I estimate the size per month as follows: First, I estimate a linear growth rate per firm reflecting the growth in Crunchbase categories per year. Therefore, I assume that all firms were in the smallest size category ("1–10"; value of 1 on a numeric scale) when they were founded. The monthly growth rate per firm f is then calculated as follows:

$$Est.\ growth\ rate\ per\ month_f$$
$$= \frac{Crunchbase\ firm\ size\ category_f - 1}{Time\ from\ founding_f\ to\ acqusition\ or\ inactive\ event,\ or\ end\ of\ observation\ period\ [month]}$$

As Crunchbase categories are pseudo-logarithmic, the linear growth factor represents an exponential growth of the firms in the observation period. Second, the estimated firm size of firm f in any given month i is then calculated as follows:

$$EstTargetSize_{fi}$$
$$= 1 + Est.\ growth\ rate\ per\ month_f * Time\ since\ founding_i\ (month)$$

This estimation assumes that Crunchbase size categories reflect the firm size either at the time of the acquisition when the firm ceased operations ("inactive") or the end of the observation period (end of 2018). Crunchbase has last updated most of the records in 2018 or 2019 and as a result, the size should approximately reflect the size at the end of the observation period. Furthermore, Hlavka (2019), who previously used Crunchbase data for acquisition timing research, confirmed that Crunchbase indeed reports correct company sizes for most acquisition targets at the time of their acquisition.

---

[22] Categories of the number of employees recorded in Crunchbase are as follows (my mapping to a numeric scale in brackets): 1 to 10 (1) 11 to 50 (2) 51 to 100 (3) 101 to 250 (4) 251 to 500 (5) 501 to 1,000 (6) 1,001 to 5,000 (7) 5,001 to 10,000 (8) 10,001 + (9) employees. As the classes are not evenly sized, this mapping leads to a pseudo-logarithmic scale.

I use a control *LnRevenue* in a robustness check, to control if revenues obtained in US\$ instead of firm size are a major determinant of a firm's attractiveness for potential acquirers. Revenues were previously used by Rogan and Sorenson (2014) as a control variable. Revenue data was collected from Orbis, which, for some firms, reports revenues on a yearly basis. Firms without revenue information were dropped from this robustness check (8,480 out of 15,386 firms considered for the "software" SIC-code 737 based sample of firms). I used log-linear interpolation to estimate missing values (e.g., in case revenues were only reported for 2016 and 2018 in Orbis, I estimated the value for 2017 based on interpolation).

In the last step of the exploratory analysis, I want to examine the role of the acquirer. Particularly, I want to examine whether there are differences across acquirers who are active in OSS themselves prior to an acquisition or not. Therefore, I created several variables concerning attributes of the acquirers. I created a dummy variable *AcquirerOSS*, reflecting whether the acquirer was active in OSS in the 12 months prior to an acquisition. The variable is created precisely the same way as the variable *FirmOSS*, which was explained at the beginning of this section. Overall, 859 out of 1,999 of the acquirers in the sample of all acquisitions and 364 out of 608 acquirers in the sub-sample of acquisitions where the target was active in OSS.

I created an independent variable *Collaboration* to explore whether direct collaboration of the acquirer and the target worked in the same OSS projects prior to the acquisition influences acquisition timing. The dummy variable takes a value of 1 if employees of target and acquirer contributed to the same projects in the year prior to the acquisition. This was only the case in 17 acquisitions. I included this variable as the relational history between the acquirer and target are known to potentially increase the odds of an acquisition, as the acquirer can obtain useful information about the target through the collaboration, or the target and the acquirer may have developed mutual trust foreshadowing a successful integration of the firms (Hernandez and Shaver, 2019; Zaheer et al., 2010).

I created a control variable *Acquisition Experience*, reflecting the acquirer's prior experience with acquisitions. The variable is a count variable, where I counted the number of acquisitions the acquirer has made in the five years prior to an acquisition based on the Crunchbase acquisition dataset. The variable was selected as it is known that acquisition experience can influence acquisition decision making (Hlavka, 2019; Rabier, 2017). For example, Hlavka (2019) showed that more acquisition experience leads to acquisitions of younger targets. The five year window was selected according to prior literature (Capron and Shen, 2007; Hlavka, 2019; Kavusan et al., 2020). 817 out of 1,999 acquirers had prior acquisitions.

I created a control variable *AcquirerSize*, which reflects the acquirer's size in Crunchbase size categories. I include this variable as an acquirer's acquisition behavior depends on the company size of the acquirer; for example, larger firms can acquire larger targets, which, in turn, are typically also older than smaller targets (Hlavka, 2019). As acquirer size is not available for all acquirers in my sample (140 acquirers missing, 147 out of 2,639 acquisitions are associated with those acquirers with missing size record), I ran regressions with a reduced sample using this variable and robustness checks with the full sample not using the variable.

Lastly, I created variables for *Country Overlap* and *Industry Overlap*. The variable *Country Overlap* captures if target and acquirer have their headquarter in the same country according to Crunchbase data. It is a binary variable taking the value 1 if the firms reside in the same country, and 0 otherwise. The variable is included, as acquiring foreign firms is more difficult due to liabilities of foreignness (Zaheer et al., 2010). Gathering full information and reducing information asymmetries prior to an acquisition is harder in non-domestic acquisitions, making it more difficult to evaluate potential targets (Graebner et al., 2010). The variable has also been used regularly in studies concerning target selection and acquisition timing (Hernandez and Shaver, 2019; Hlavka, 2019). 33 acquisitions do not have a record for the *Country Overlap* variable as the target or the acquirer had no record for their country. Those are dropped in regressions where this variable is used. The variable *Industry Overlap* captures if target and acquirer work in a similar industry according to Crunchbase data. I include this variable, as acquirers might be better at identifying and evaluating acquisition opportunities in their own field (Haleblian and Finkelstein, 1999; Hlavka, 2019; Pennings, Barkema, and Douma, 1994). In line with Hlavka (2019), I do a paired comparison of the list of Crunchbase's acquirer industry categories and target industry categories to create the variable *Industry Overlap*. If at least one category is identical, the binary variable is set to 1, and 0 otherwise.

4.4 Results

4.4.1 Descriptive Results

This section gives an overview of the characteristics of the data sample utilized in this chapter. Therefore, I provide three tables showing the univariate descriptive statistics and correlations in line with the three steps of the analysis: A table showing the descriptive statistics for the full sample of firms active in OSS and firms not active in OSS for the analysis of firms active in OSS compared to firms not active in OSS at the end of the observation period (Table 4.3; see digital Appendix E-1 for firm-month level descriptive statistics instead of firm level descriptive statistics); a table showing the descriptive statistics only for the set of firms active in OSS development for the analysis of characteristics of a firm's OSS development (Table 4.4; see digital Appendix E-2 for firm-month level instead of firm level descriptive statistics); and a table showing the descriptive statistics at the end of the observation period per acquisition target, including all variables concerning the acquirer for an analysis of the role of the acquirer's OSS activity (Table 4.5). For reasons of clarity, each table contains only the variables required for the models in which the respective data is used. Additionally, Table 4.6 shows a comparison of means, standard deviations (S.D.), minimum values, and maximum values of the full sample of firms active in OSS development and the firms not active in OSS development.

Looking at the full panel dataset (Table 4.3; OSS-active and non-active firms), it becomes clear that the dataset mostly captures the early years of a company's life. On average, I have 51.8 monthly observations per company, which is equal to the first 4.3 years of a company's life. Several firms are observable for the full eight years of the observation period (96 monthly observations). The estimated target size variable, along with all other independent and control variables, is positively correlated with acquisition event. This is also true for the independent variables describing the characteristics of firms' activities in OSS (Table 4.4). The positive correlation of the independent and control variables with the occurrence of acquisition events is also reflected in the descriptive statistics of the sample of realized acquisitions (Table 4.5), which capture the characteristics of the firms at the time of their acquisition. Firms in the sample get acquired after 43 months on average (~3.6 years). 31% of the acquired firms had been active in OSS in the year prior to their acquisition, 50% of them had previously received investments, and, on average, firms had filed for 0.31 patents ($\ln(1 + 0.31) = ~0.27$).

Turning to differences of the group of firms active in OSS development and firms never active in OSS development (Table 4.6), firms active in OSS development are on average 10.3 months older than firms not active in OSS development. This discrepancy is driven by the fact that firms founded late in the observation period are rarely active in OSS. Firms first need to start developing software—be it their own product or contributions to other OSS projects—before they can make contributions to OSS (see digital Appendix E-3 for a yearly comparison of firms founded and acquired thereof). 8% of the firms active in OSS development got acquired in the observation period compared to 4% not active in OSS development. In line with the positive correlation of the *FirmOSS* variable with the controls *HasInvestors, LnPatents, and EstFirmSize* (Table 4.3), the means for those variables are higher for the firms active in OSS development compared to the firms not active in OSS development (*HasInvestors*: 0.52 vs. 0.26; *LnPatents*: 0.31 vs. 0.10; *EstFirmSize*: 1.97 vs. 1.43). t-tests comparing the means of both groups show that the means of the variables are significantly different for the groups. Therefore, besides the control variables, I employ matching for one of my robustness checks, to account for potential differences of the groups.

Turning to acquirer characteristics (included in Table 4.5), in 56% of the acquisitions in the sample, the acquirer was active in OSS. Most acquisitions were domestic acquisitions (69%), and 46% of the acquirers shared at least one Crunchbase industry category with the target. On average, the acquirers had conducted 1.72 acquisitions in the five years prior to the focal acquisition. The variable is heavily right-skewed with some serial acquirers, mostly tech giants, like Google and Microsoft, having conducted up to 57 acquisitions in five years.

4.4.2   Exploring Acquisition Likelihood of OSS-active and Non-active Firms

In this section, I analyze differences between firms active in OSS development and firms not active in OSS development and their likelihood of getting acquired. Therefore, I compare the acquisition hazard for firms active in OSS development compared to firms not active in OSS development using a Cox proportional hazard model (CPHM) and controlling for several control variables (Cox, 1972). Such models are commonly used to identify factors influencing acquisition likelihood in management research (e.g., Fischer et al., 2020; Ransbotham and Mitra, 2010). In the hazard model, the dependent variable (DV) is an acquisition event. The CPHM estimates the baseline hazard and a coefficient for each explanatory covariate describing how the baseline hazard changes in response to changes in the covariates. In the CPHM, the baseline hazard describes how the risk of a so-called "failure event" changes over time when covariates are at the mean level, and the parameters are assumed to have a multiplicative effect on the baseline hazard (Bradburn, Clark, and Love, 2003; Cox, 1972; Ransbotham and Mitra, 2010). The "failure event" is defined here as an acquisition. As typical for models of this type, an acquisition event may or may not happen before the end of the observation period. Hence, acquired and non-acquired firms are considered in the model.

Table 4.7 shows the results of the analysis. Model 1 is a CPHM only run with the main controls *HasInvestors, LnPatents, and EstFirmSize*. Model 2 is run with

Table 4.3 Descriptive statistics and correlations for all firms at the end of the observation period (incl. alternative events)

Variables		Descriptive statistics				Pairwise correlation coefficients								
		Mean	S.D.	Min	Max	[1]	[2]	[3]	[4]	[5]	[6]	[7]	[8]	[9]
[1]	Firm age (months)	50.84	25.78	1	95	1.00								
[2]	Event: Acquisition	0.04	0.19	0	1	-0.06 (0.00)	1.00							
[3]	Event: Ceased operations	0.04	0.19	0	1	-0.13 (0.00)	-0.04 (0.00)	1.00						
[4]	Event: IPO	0.002	0.04	0	1	-0.02 (0.00)	-0.01 (0.04)	-0.01 (0.04)	1.00					
[5]	FirmOSS	0.13	0.34	0	1	0.08 (0.00)	0.10 (0.00)	-0.02 (0.18)	0.01 (0.00)	1.00				
[6]	HasInvestors	0.31	0.46	0	1	0.05 (0.00)	0.08 (0.00)	0.01 (0.02)	0.03 (0.00)	0.21 (0.00)	1.00			
[7]	LnPatents	0.14	0.53	0	8.09	0.11 (0.00)	0.05 (0.00)	-0.02 (0.72)	0.02 (0.00)	0.14 (0.00)	0.18 (0.00)	1.00		
[8]	EstFirmSize	1.53	0.84	1	9	0.16 (0.00)	0.09 (0.00)	-0.06 (0.00)	0.03 (0.00)	0.26 (0.00)	0.17 (0.00)	0.14 (0.00)	1.00	
[9]	LnRevenue (n = 8,480 firms from Orbis)	11.00	2.43	0.48	23.15	0.53 (0.00)	-0.16 (0.00)	-0.09 (0.51)	0.05 (0.00)	0.18 (0.00)	0.02 (0.00)	0.05 (0.00)	0.32 (0.00)	1.00

Note: N = 67,794 firms; p-values in parentheses; all variables are time-varying except HasInvestors.

Table 4.4 Descriptive statistics and correlations for firms active in OSS development at the end of the observation period

Variables		Descriptive statistics				Pairwise correlation coefficients							
		Mean	S.D.	Min	Max	[1]	[2]	[3]	[4]	[5]	[6]	[7]	[8]
[1]	Firm age (months)	57.06	223.11	2	95	1.00							
[2]	Event: Acquisition	0.10	0.30	0	1	−0.14 (0.00)	1.00						
[3]	LnCommits	1.26	1.97	0.69	11.18	−0.11 (0.00)	0.07 (0.00)	1.00					
[4]	PullRequests	0.36	0.48	0	1	0.02 (0.02)	0.05 (0.00)	0.49 (0.00)	1.00				
[5]	HasInvestors	0.52	0.50	0	1	0.01 (0.57)	0.09 (0.00)	0.11 (0.00)	0.11 (0.00)	1.00			
[6]	LnPatents	0.31	0.81	0	7.83	0.08 (0.00)	0.04 (0.00)	0.06 (0.00)	0.06 (0.00)	0.20 (0.00)	1.00		
[7]	EstFirmSize	1.97	1.06	1	9.0	0.16 (0.00)	0.05 (0.00)	0.22 (0.00)	0.16 (0.00)	0.19 (0.00)	0.17 (0.00)	1.00	
[8]	License (0: Copyleft; 1: Permissive)	0.87	0.34	0	1	−0.004 (0.74)	0.004 (0.72)	−0.07 (0.00)	−0.02 (0.09)	0.05 (0.00)	0.02 (0.15)	0.04 (0.00)	1.00

Note: N = 12,597 firms (not including IPOs and firms that ceased operations); p-values in parentheses. All variables are time-varying except HasInvestors. Only firms required for survival analysis (i.e., firms which went public or ceased operations are not included).

Table 4.5 Descriptive statistics and correlations for target and acquirer characteristics at the time of realized acquisitions

Variables		Descriptive statistics				Pairwise correlation coefficients										
		Mean	S.D.	Min	Max	[1]	[2]	[3]	[4]	[5]	[6]	[7]	[8]	[9]	[10]	[11]
[1]	Firm age (months; at acquisition)	43.05	19.95	1	95	1.00										
[2]	FirmOSS	0.31	0.46	0	1	0.15 (0.00)	1.00									
[3]	HasInvestors	0.50	0.50	0	1	0.19 (0.00)	0.22 (0.00)	1.00								
[4]	LnPatents	0.27	0.72	0	6.01	0.16 (0.00)	0.15 (0.00)	0.18 (0.00)	1.00							
[5]	EstFirmSize	1.94	1.24	1	9	0.16 (0.00)	0.15 (0.00)	0.08 (0.00)	0.16 (0.00)	1.00						
[6]	AcquirerOSS	0.54	0.50	0	1	0.05 (0.02)	0.21 (0.00)	0.14 (0.00)	0.08 (0.00)	0.01 (0.61)	1.00					
[7]	Collaboration (in OSS)	0.02	0.14	0	1	0.05 (0.01)	0.19 (0.00)	0.06 (0.03)	0.09 (0.00)	0.16 (0.00)	0.08 (0.00)	1.00				
[8]	AcquisitionExperience	1.73	5.30	0	57	0.005 (0.82)	0.13 (0.00)	0.09 (0.00)	0.10 (0.00)	-0.01 (0.58)	0.23 (0.00)	0.08 (0.00)	1.00			
[9]	AcquirerSize (Crunchbase categories)	5.12	2.56	1	9	0.08 (0.00)	0.22 (0.00)	0.16 (0.00)	0.23 (0.00)	0.19 (0.00)	0.41 (0.00)	0.12 (0.00)	0.35 (0.00)	1.00		
[10]	SameCountry	0.69	0.46	0	1	-0.05 (0.01)	0.12 (0.71)	0.12 (0.00)	0.02 (0.21)	0.03 (0.17)	0.05 (0.01)	0.04 (0.05)	0.06 (0.00)	-0.00 (0.88)	1.00	
[11]	SameIndustry	0.46	0.50	0	1	0.001 (0.99)	0.05 (0.01)	0.09 (0.00)	0.02 (0.31)	0.02 (0.36)	0.08 (0.00)	-0.002 (0.97)	-0.05 (0.03)	-0.02 (0.53)	0.04 (0.06)	1.00

Note: N = 2,472 acquisitions (observations with missing records of additional variables compared to previous tables dropped here); p-values in parentheses.

Table 4.6 Comparison of descriptive statistics of firms active in OSS development and firms not active

Variables		Firms active in OSS (n = 13,036 firms)				Firms not active in OSS (n = 54,758 firms)				t-test means
		Mean	S.D.	Min	Max	Mean	S.D.	Min	Max	p-value
[1]	Firm age (months)	59.16	22.57	4	95	48,86	26.10	1	95	<0.01
[2]	Event: Acquisition	0.08	0.27	0	1	0.04	0.17	0	1	<0.01
[3]	Event: Ceased operations	0.03	0.17	0	1	0.04	0.19	0	1	<0.01
[4]	Event: IPO	0.004	0.06	0	1	0.001	0.03	0	1	<0.01
[5]	HasInvestors	0.52	0.50	0	1	0.26	0.44	0	1	<0.01
[6]	LnPatents	0.31	0.81	0	7.83	0.10	0.43	0	8.09	<0.01
[7]	EstFirmSize	1.97	1.06	1	9	1.43	0.74	1	9	<0.01
[8]	LnRevenue (n = 8,480 firms from Orbis)	11.55	2.40	1.47	23.15	10.53	2.35	0.48	23.15	<0.01

Note: Including firms that ceased operations or went public; two-sided t-tests.

the independent variable *FirmOSS* as the only regressor. Model 3 shows the full model. Models 4–7 show robustness checks: Model 4 utilizes the ln of a firm's commits to OSS in the 12 months prior to the focal month (+1) as an alternative independent variable. Model 5 utilizes a matched sub-sample of the main sample, where OSS-active and non-active firms were matched on three key characteristics of the firms: country, year of founding, and Crunchbase industry tags. For the matching of category tags, I did pairwise comparison of the firms' tags, and a match on categories was created if the firms had at least two tags in common. I use this highly granular matching to ensure that firms are active in a similar field within the software industry. Overall, the matched sample contains 10,300 firms active in OSS development and 32,776 not active. For this regression, I employ weights to account for a varying number of matches per firm.[23] Model 6 and 7 utilize the alternative sample, where I utilized SIC-codes sourced from Orbis to identify firms in the software industry. Model 6 utilizes the identical variables as the full model, while Model 7 uses the yearly *Revenue* as an alternative control instead of *EstFirmSize*.

Furthermore, I ran a competing risk model, which also accounts for alternative events next to acquisitions using the same variables as the full model. The competing risk model is shown in Table 4.8.

In all models presented in this section a coefficient $\beta > 0$ indicates that an increase in the associated covariate is associated with an increasing acquisition hazard. In turn, a negative coefficient indicates that an increase in the associated covariate is associated with a decreasing acquisition hazard.

Across all models, the results show that a firm's being active in OSS development is associated with a higher acquisition hazard (i.e., such firms are more likely to be acquired; $\beta = 0.80$, $p<0.01$ in main model M3). The controls *LnPatents, HasInvestors*, and *EstFirmSize* are also associated with a higher acquisition hazard ($\beta = 0.44$, $p<0.01$; $\beta = 0.08$, $p<0.01$; $\beta = 0.54$, $p<0.01$). Patents are known to increase acquisition likelihood (Ransbotham and Mitra, 2010). Hence, this finding is in line with prior literature. Investors might be associated with a higher acquisition hazard, as investors, particularly VCs focusing on young firms,

---

[23] I applied the weighting logic known from Coarsened Exact Matching (Blackwell, Iacus, and Porro, 2009), where I fixed the weight of all firms active in OSS development to 1 and adjusted the weights of the non-active firms accordingly. E.g., if one "OSS-active" firm matches with two "non-active firms", each non-active firm gets the weight 0.5. If one "non-active firm" matches with two "OSS-active" firms, this non-active firm gets a weight of 2.

often seek an exit by selling their investment to an acquirer.[24] Lastly, the positive coefficient associated with *EstFirmSize* suggests that acquirers prefer firms that have proven their potential for (future) success, reflected in their increasing size.

Figure 4.2 graphically emphasizes the increased acquisition hazard associated with firms active in OSS development; the graph on the right shows the effect adjusted for the controls.[25] The difference between the curve for the firms active in OSS development and the curve for the firms not active is smaller in the right graph than the left graph as all controls are positively correlated with the independent variable and have a positive regression coefficient.

The competing risk model (Table 4.8) shows similar results for acquisition events supporting the results from the main survival model. Furthermore, it shows that changes in the likelihood to conduct an IPO and the rate at which firms cease operations are not associated with changes in a firm's activity in OSS ($\beta = 0.22$, p $= 0.36$ and $\beta = 0.08$, p $= 0.29$, respectively). Moreover, it shows that larger size is associated with a lower likelihood to cease operations ($\beta = -0.30$, p<0.01, respectively). However, having investors is associated with a higher likelihood to cease operations ($\beta = 0.13$, p<0.01). This might be explained by investors stepping in earlier to stop unsuccessful ventures they have invested in instead of continuing a probably unsuccessful venture, like the founders of the venture might do. For IPOs, I find that firm size and having investors is associated with a higher likelihood to conduct an IPO ($\beta = 0.64$, p<0.01; and $\beta = 1.26$, p<0.01, respectively). Having a higher number of filed patents is weakly associated with a higher likelihood to go public ($\beta = 0.19$, p $= 0.09$). Results regarding IPOs are robust to including year dummies to account for hypes in IPO activity (not reported).

---

[24] Hlavka (2019) found that if a target received investments, it is on average acquired earlier than if it did not, which supports the argument that investors may seek an earlier exit which in turn increases acquisition likelihood for young firms.

[25] Adjusting for controls is a function part of Stata's *ststs*-package. When adjusting for controls, Stata fits a separate CPHM on the independent and control variables and retrieves the separately estimated baseline survivor function per variable (see https://www.stata.com/manuals13/ststs.pdf; last accessed 12.01.2021). Without adjusting for controls, the graphs show the share of firms from either group that at an age of t month had not yet been acquired as estimated by the full CPHM.

Table 4.7 Regression results CPHM comparing firms active in OSS with non-active firms

Variables	[1] Controls only	[2] Dummy OSS w/o controls	[3] Dummy OSS and controls	[4] LnCommits (+1)	[5] Matched (weighted)	[6] Orbis SIC-code sample	[7] Revenue as control.
FirmOSS	-	1.25*** (0.04)	0.80*** (0.05)	-	0.81*** (0.06)	0.44*** (0.09)	0.34*** (0.11)
LnCommits (+1)	-	-	-	0.14*** (0.01)	-	-	-
HasInvestors	0.58*** (0.04)	-	0.44*** (0.04)	0.49*** (0.04)	0.47*** (0.06)	0.07 (0.08)	0.07 (0.11)
LnPatents	0.13*** (0.02)	-	0.08*** (0.03)	0.10*** (0.03)	0.06** (0.03)	0.13*** (0.04)	0.10* (0.05)
EstFirmSize	0.59*** (0.02)	-	0.54*** (0.02)	0.55*** (0.02)	0.44*** (0.03)	0.52*** (0.03)	0.51*** (0.05)
LnRevenue (Orbis subset only)	-	-	-	-	-	-	-0.001 (0.02)
Concordance	0.68	0.59	0.69	0.68	0.70	0.69	0.67
Wald χ^2	2,036***	853***	2,421***	2,346***	1,035***	446***	249***
Log-likelihood	-26,943	-27,276	-26,805	-26,860	-12,783	-5,991	-3,474
N firms	65,262	65,262	65,262	65,262	43,076	15,386	3,477
N events	2,639	2,639	2,639	2,639	2,077	684	439

Note: Cox proportional hazard estimates on 3,427,814 monthly observations for 65,262 firms (base case); failure is acquisition; * $p < 0.10$, ** $p < 0.05$, *** $p < 0.01$ (two-tailed significance); standard errors in parentheses. All covariates except HasInvestors are time-varying covariates.

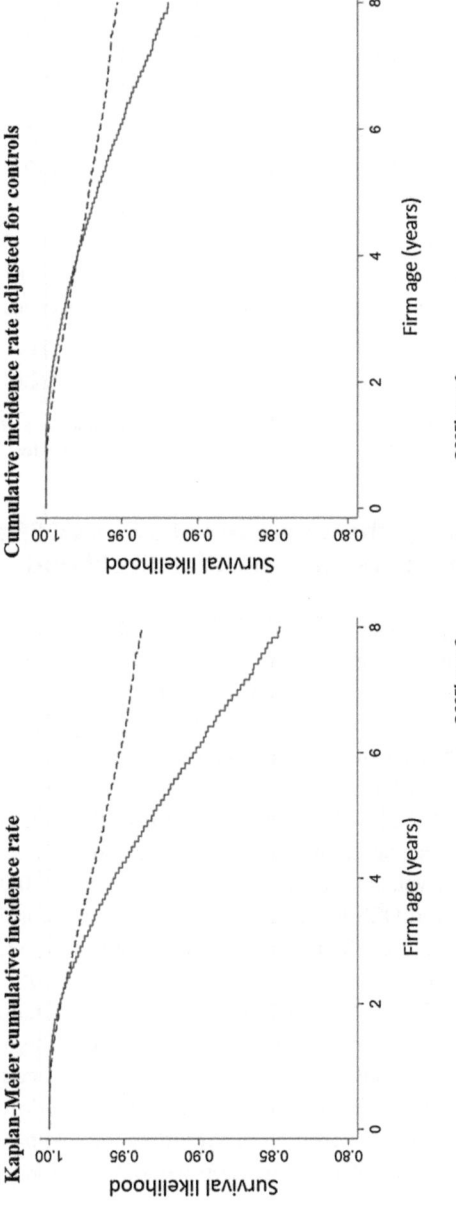

Figure 4.2 Kaplan–Meier cumulative incidence rate (main model)

Table 4.8 Regression results of competing risk model

Variables	[1] Focal event: Acquisition	[2] Focal event: Ceased operations	[3] Focal event: IPO
FirmOSS	0.81*** (0.05)	0.08 (0.07)	0.22 (0.24)
HasInvestors	0.43*** (0.04)	0.13*** (0.04)	1.26*** (0.22)
LnPatents	0.09*** (0.03)	−0.03*** (0.05)	0.19* (0.11)
EstFirmSize	0.54*** (0.02)	−0.30** (0.06)	0.64*** (0.05)
Wald χ^2	2,475***	30.58**	348.5***
Log-likelihood	−26,936	−25,879	−1,104
N firms	67,794	67,794	67,794
N events	2,639	2,421	111
N competing events	2,532	2,750	5,060

Note: * $p<0.10$, ** $p<0.05$, *** $p<0.01$ (two-tailed significance); standard errors in parentheses. All covariates except HasInvestors are time-varying covariates.

4.4.3   Exploring three characteristics of a firm's OSS activities and their influence on acquisition likelihood

In the second step of my analysis, I want to uncover differences within the group of firms active in OSS development. Again, using a CPHM and using the same DV (acquisition event), I focus on three key characteristics of their OSS engagement as independent variables: their overall level of OSS activity (Table 4.9, Models 1–3), their ability to attract external contributions (Models 4–7), and the OSS licenses the firms employ (Models 8, 9).

Regarding the role of the overall level of a firm's OSS activity (Table 4.9, Model 2), I find that increasing commit activity of the firm comes with an increasing acquisition hazard, yet, this result is not significant ($\beta = 0.03$, $p = 0.13$). When using the categorized OSS levels instead of the log-transformed number of contributions, the regression coefficient indicates a positive association between acquisition hazard and increasing levels of a firm's activity in OSS development and is significant (Model 3; $\beta = 0.12$, $p<0.01$). A closer examination using dummies for each category reveals that firms with a medium or a high activity level in OSS development are associated with a significantly higher acquisition hazard than firms with a low activity level. However, no significant difference can be found between medium to high levels of OSS activity (digital Appendix F). Model 1 (Table 4.9) shows the results of the regression without controls.

Table 4.9 Regression results CPHM with focus on characteristics of a firm's OSS activities

Variables	Level of OSS activity			Ability to attract outside contributions				License choice	
	[1] Commit activity variable only	[2] With controls	[3] Alternative specification	[4] PullReq variable only (dummy)	[5] With controls	[6] Alt. specification I Pull Requests (ln)	[7] Alt. specification II nr ext. contributors	[8] License variable only	[9] With controls
Commits (ln)	0.06*** (0.02)	0.03 (0.02)	–	–	–	–	–	–	–
Commits categorized	–	–	0.12*** (0.04)	–	–	–	–	–	–
Pull Requests (dummy)	–	–	–	0.23*** (0.07)	0.09 (0.07)	–	–	–	–
Ln 12 m pull requests	–	–	–	–	–	–0.02 (0.02)	–	–	–
Nr. External contributors	–	–	–	–	–	–	0.0001 (0.01)	–	–
License (0: Copyleft; 1: Permissive)	–	–	–	–	–	–	–	0.15 (0.12)	0.06 (0.12)
HasInvestors	–	0.36*** (0.08)	0.36*** (0.08)	–	0.35*** (0.08)	0.37*** (0.08)	0.36*** (0.08)	–	0.32*** (0.09)
LnPatents	–	0.06* (0.04)	0.06* (0.03)	–	0.06* (0.04)	0.06* (0.03)	0.06* (0.03)	–	0.06* (0.04)
EstFirmSize	–	0.36*** (0.04)	0.36*** (0.04)	–	0.36*** (0.04)	0.37*** (0.04)	0.37*** (0.04)	–	0.34*** (0.04)
Concordance	0.54	0.61	0.61	0.53	0.62	0.61	0.62	0.51	0.61
Wald χ^2	11.9***	186***	190***	10.3***	181***	182***	180***	1.67	123***
Log-likelihood	–6,638	–6,559	–6,556	–6,638	–6,559	–6,559	–6,560	–5,366	–5,307
N firms	12,597	12,597	12,597	12,597	12,597	12,597	12,597	10,047	10,047
N events	808	808	808	808	808	808	808	674	674

Note: * $p<0.10$, ** $p<0.05$, *** $p<0.01$; standard errors in parentheses. All covariates except HasInvestors are time-varying covariates.

Regarding the role of a firm's ability to attract outside contributions, I do not find significant results (Models 4–7). While the regression without controls suggests an increasing acquisition hazard with increasing outside contributions (Model 4; $\beta = 0.23$, p<0.01), the result is not robust to including controls (Model 5; $\beta = 0.09$, p $= 0.21$). I get similar results when replacing the independent dummy variable *PullRequests* with either the log-transformed number of pull requests (Model 6; $\beta = -0.02$, p $= 0.24$) or the number of external contributors creating pull requests to a firm's projects (Model 7; β<0.0001, p $= 0.92$).

Regarding the role of the licenses a firm focuses on, I also do not find significant results (Models 8, 9). The license type is insignificant, both in Model 1 without controls ($\beta = 0.15$, p $= 0.20$) and in Model 2 with controls ($\beta = 0.086$ p $= 0.58$).

4.4.4   The Role of an Acquirer's OSS Engagement for Timing of Acquisitions

In the last step of my analysis, I explore the role of the acquirer. To do so, I focus on the timing of acquisitions and examine whether acquirers active in OSS themselves buy targets at a younger or older age than acquirers not active in OSS using ordinary least squares (OLS) models.[26]

I utilize different model specifications to examine the role of an acquirer's activity in OSS prior to the acquisition of a firm. The results of those are shown in Table 4.10. Model 1 shows the results for the OLS regression only conside- ring the controls. Model 2 shows the results for the sample of acquisition targets active in OSS development and targets not active in OSS and an interaction term to examine the role of an acquirer's OSS activity when acquiring targets active

---

[26] I choose this OLS model in line with previous research on acquisition timing (e.g., Fischer et al., 2019; Hlavka, 2019; Stein, 2017), including research on timing and the role of cha- racteristics of the acquirer (Hlavka, 2019). I do not use a CPHM in this section, as CPHM does not allow to examine acquirer characteristics (surviving firms don't have an acquirer). On the other hand, I do not use the OLS model in the two previous sections to examine target characteristics, as my data is limited to the early years of a firm's life and a large share of firms does not get acquired in the observation period, but might get acquired later (i.e., given the higher acquisition hazard for firms active in OSS development compared to non-active firms, there are more non-active firms "surviving" my observation period which could get acquired in the future). These potentially missing acquisitions are no problem when examining cha- racteristics of the acquirer *within* the group OSS-active targets, which are in the focus of this section. Obviously, one needs to keep in mind, that the results obtained are only valid for young firms, which are in the focus of this chapter.

in OSS. In this model, I interact the dummy variable *FirmOSS*, which captures a target's involvement in OSS development, and the dummy variable *AcquirerOSS*, which captures an acquirer's involvement in OSS development. The interaction term then captures the average difference in acquisition timing between the acquirers active in OSS and the acquirers not active in OSS for targets active in OSS. To be able to understand the role of an acquirer's involvement in OSS development for acquisition timing even better, I additionally show a model explicitly differentiating the four combinations of the two dummy variables for a target's and an acquirer's involvement in OSS (Model 3). The base case is both firms not being active in OSS. Model 4 then shows the sub-sample only focusing on acquisitions, where the target has been active in OSS development prior to their acquisition as a robustness check.

All models confirm that acquirers active in OSS development acquire firms active in OSS development at a younger age than acquirers not active in OSS development. Specifically, they acquire them 2.99 months earlier than acquirers not active in OSS development (Model 2, $\beta_{TargetOSS, AcquirerOSS}-\beta_{TargetOSS, AcquirerNotOSS} = 3.29{-}6.28$: the significance of the difference of the coefficients is confirmed with a Wald-test, $p = 0.06$). There is no significant difference in the timing of acquisitions of targets not active in OSS development between acquirers active in OSS development and acquirers not active in OSS development ($\beta_{TargetNotOSS, AcquirerOSS} = 1.49$, $p = 0.14$). I do not find evidence that direct collaboration in OSS between a target and its acquirer is associated with delayed acquisitions (Model 2; $\beta = 3.53$, $p = 0.45$).

Lastly, one might wonder if the positive association between an acquired target's involvement in OSS development and the age at which it is acquired (regression coefficient for FirmOSS in Model 2 is 6.28, $p < 0.01$) contradicts the findings from section 4.4.2 that a firm's being involved in OSS is associated with a higher acquisition hazard. This is not the case. A higher acquisition hazard in one group (A) compared to another group (B) implies an earlier mean age at acquisition in Group A only if the share of acquired firms is the same in both groups in the period of observation. This, however, is not the case here—the share of acquired firms is larger among the firms active in OSS. Scaling the curves in the Kaplan-Meier graph (see section 4.4.2) such that both the starting points and the endpoints coincide shows that the curve representing the OSS firms lies above the curve for the non-OSS-firms since it is more concave. Hence, the average age of acquisition, conditional on being acquired, is higher for OSS firms. A longer observation period, ideally covering most firms' entire life cycle, might reverse this result.

Table 4.10 OLS regression results examining the role of the acquirer

DV: target age in months at acquisition Variables	[1] Controls only	[2] Interaction	[3] Categorized combinations of OSS activity	[4] Targets active in OSS only
FirmOSS (target)	–	6.28*** (1.47)	–	–
AcquirerOSS	–	1.49 (1.00)	–	−2.93* (1.62)
FirmOSS × AcquirerOSS	–	−4.48** (1.81)	–	–
FirmOSS = 0 AcquirerOSS = 1	–	–	1.49 (0.98)	–
FirmOSS = 1 AcquirerOSS = 0	–	–	6.28*** (1.47)	–
FirmOSS = 1 AcquirerOSS = 1	–	–	3.29*** (1.16)	–
Collaboration	–	3.53 (4.72)	3.53 (4.72)	3.33 (4.70)
Acquisition Experience	−0.15 (0.08)	−0.07 (0.08)	−0.07 (0.08)	0.02 (0.11)
AcquirerSize	0.09 (0.17)	−0.23 (0.18)	−0.23 (0.18)	−0.55* (0.33)
HasInvestors (target)	6.88*** (0.81)	6.36*** (0.82)	6.36*** (0.82)	7.89*** (1.51)
LnPatents (target)	3.23*** (0.56)	3.12*** (0.56)	3.12*** (0.56)	1.93** (0.80)
EstFirmSize (target)	2.32*** (0.33)	2.20*** (0.34)	2.20*** (0.34)	2.51*** (0.54)

(continued)

Table 4.10 (continued)

DV: target age in months at acquisition Variables	[1] Controls only	[2] Interaction	[3] Categorized combinations of OSS activity	[4] Targets active in OSS only
SameCountry	−3.30*** (0.85)	−3.26*** (0.85)	−3.26*** (0.85)	−6.88*** (1.51)
SameIndustry	−0.60 (0.78)	−0.74 (0.78)	−0.74 (0.78)	−1.29 (1.38)
Constant	37.56*** (1.22)	37.04*** (1.24)	37.04*** (1.24)	46.21*** (2.48)
N	2411	2411	2411	752
R^2	0.08	0.09	0.09	0.10
Adjusted R^2	0.08	0.08	0.08	0.09
F-statistic	29.92***	21.19***	21.19***	9.43***

Note: * p<0.10, ** p<0.05, *** p<0.01; standard errors in parentheses.

4.5   Summary of Quantitative Results and Discussion of Mechanisms

In this section, I briefly recapitulate the main findings from the quantitative analysis and then discuss potential mechanisms behind these findings in light of theory and findings from the qualitative interviews.

The first key finding from my exploratory analysis is that being active in OSS development is associated with a higher acquisition hazard for young firms active in software development. While I cannot claim to fully prove causality, several robustness checks, as well as a competing risk model, confirmed the association. I see five potential mechanisms that might be the driver of this result. In the following, I want to discuss those and set them in the context with prior literature and the interviews to evaluate their relevance as drivers of the effect.

First, being active in OSS might increase the visibility of a firm towards potential acquirers. I.e., acquirers more often find targets active in OSS development compared to firms not active in OSS development during their target search process. This mechanism would be in line with the argument that OSS increases visibility for firms towards potential new hires (Henkel, 2004) or for potential customers (Henkel, 2006) and findings from my interviews (see Section 3.3.2) where, for example, one senior manager of an acquirer mentioned: *"I mean, had they not been doing it [target search] in Open Source, I don't think that acquisition would have ever happened. [...] The technology was so early. [...] So, it was really like, where do you find experts in this tech? You go to the communities that are building them, and so I think that was it [how acquirer found the target]."*

Second, OSS development activities of a firm are better to evaluate. OSS might thus allow an acquirer to reduce uncertainty around a target's technology, which would make firms active in OSS development more attractive as targets than firms not active in OSS development. This should particularly be true in firms that focus on software development, such as those in my sample. This mechanism is also supported by my findings from the interviews (see Section 3.3.2), where, for example, one CEO of an acquirer explained: *"the reason Open Source is interesting is, that you can, unlike acquiring a proprietary company or a company with proprietary software, you can see the quality of the engineers before you buy, because you can see their Open Source contributions, you can go through their repos and see how they think about products and architecture."* Additionally, the second step of my quantitative analysis (Section 4.4.3) provides further insights regarding this argument. On the one hand, firms being more often acquired when they are more active in OSS supports the argument at hand: particularly if firms facilitate a larger share of their software development in OSS, there would be more to observe and evaluate, which, in turn, can reduce uncertainty around a target. On the other hand, I do not find evidence that being successful in receiving contributions from an external community would increase acquisition hazard. The number of pull-requests received did not influence acquisition hazard. While interviewees highlighted the role of examining the community and their contributions—for example, one acquirer said *"we look at their contributions, how many contributions they had; all that sort of stuff [to evaluate a target]"*—I do not find evidence this would quantitatively influence acquisition hazard. Hence, while selected OSS activity of a potential target might increase its attractiveness due to the observability of the activity, not all sorts of activities in OSS do so the same way. The analyses presented in this chapter suggest that, for example, acquirers might use a target's OSS activities rather to evaluate the quality of the target's engineers instead of the target's ability to attract a large outside community.

Third, participating in OSS may allow firms to build better products faster, making them attractive targets. Particularly young firms engage in OSS development to overcome liabilities of newness and smallness, such as a lack of resources (Gruber and Henkel, 2006). For example, young firms participating in OSS can utilize input from a community as an external resource to the firm (Dahlander and Magnusson 2008), and they might be able to build product features that would have been hard to envision if applying a proprietary approach (Almirall and Casadesus–Masanell, 2010; Henkel and von Hippel 2005; Hepp, 2016). These advantages might lead to building better products, potentially also faster compared to firms developing everything internally without interacting in OSS. Regarding this mechanism, I did not find any evidence in the interviews; comments were neither in line with the argument nor against it. Also, the quantitative analysis did not provide much guidance. The competing risk model showed that firms active in OSS get acquired more often, but there was no significant difference between firms active in OSS and not active in OSS and their likelihood to stop operations. Hence, being active in OSS does not "protect" a firm from failing. As a result, I do not have clear evidence that being active in OSS generally helps firms outperform firms not active in OSS. More research is required to examine whether the suggested mechanism plays any role in acquisition likelihood and timing.

Fourth, being active in OSS might allow the firms to increase their user base more quickly than firms focusing on proprietary software. A large user base was regularly mentioned as an acquisition motive for firms active in OSS development. As one CTO and Co-founder explained: *"We wanted [to acquire] something that had a wide adoption. And the only way to get wide adoption in that space was to be Open Source."* From an analysis perspective, however, more users should, on average, lead to more contributions, as some of the users might make changes to the software which they want to contribute back to the main repository. Yet, the analysis of pull requests to a firm's repositories was not associated with changes in acquisition likelihood in step two of my analysis. Hence, as I do not have quantitative data on the usage on the OSS developed by the firms in my sample, further analysis would be required to assess this potential mechanism.

Fifth, OSS might be a hyped technology trend, and, thus, more acquisitions of OSS companies take place in my observation period. Hlavka (2019) has shown that technology hypes can lead to earlier acquisitions of firms focussing on that technology. However, OSS was identified as a non-hyped technology in the observation period of Hlavka (2019), which covered the timeframe from 2002 to 2017 and, hence, has a large overlap with my dataset. Furthermore, I did not find evidence that OSS was particularly hyped in my interviews. Therefore, hype as the main driver of the effect observed in my analysis can be excluded.

Sixth and lastly, one could argue that owners of firms active in OSS more often sell their company because they struggle to monetize the OSS. Research has shown that monetizing software that has been open sourced is difficult and requires creative business models to allow profiting from OSS (Dahlander and Magnusson, 2008; West and Gallagher, 2006), something the firms may fail to do successfully, forcing owners to seek an exit before filing for bankruptcy. However, my analysis shows that this should not be a main driver of the effect observed as there is no significant difference between firms active in OSS and firms not active in OSS with regard to the number of firms ceasing operations. This should be the case if firms active in OSS would be less successful.

Taken together, four potential mechanisms, or a combination thereof, could drive the observed effect. Firms active in OSS could be more visible to acquirers and more attractive as they are better to evaluate and potentially able to build better products and attract a larger user base. I do not have evidence that a technology hype of OSS would be a driver nor that firms active in OSS would be sold more often due to not being, on average, less successful than firms not active in OSS.

In the second step of my analysis, I focused on the group of firms active in OSS development. I found—while not significant across all models—that medium to high levels of activities are associated with a higher acquisition hazard than firms with only low levels of OSS activity, while I did not find evidence that a firm's ability to attract outside contributions or the licenses they focus on significantly influence acquisition likelihood. As already mentioned above, these results give an indication for what the acquirers focus on. Acquirers seem to focus rather on a target's (contributions to) software development instead of the community's contributions to the target's projects or the type of licenses a target focuses on.

Hence, they rather seem to evaluate the engineers of a target, which would be an indicator for the motive of many of the acquisitions at hand: Acquirers seem to rather acquire the engineering talent a target possesses in the style of an acqui-hire (Coyle and Polsky, 2013) than the OSS. This argument is supported by the large number of acquisitions in my interviews where *OSS talent* was mentioned as an acquisition motive (see Section 3.3.1).

In the last step of my analysis, focussing on the role of an acquirer's activity in OSS, I find that acquirers active in OSS themselves acquire firms active in OSS earlier than acquirers not active in OSS. I see two potential mechanisms driving this result. First, acquirers themselves active in OSS might have earlier access to potential targets than firms not active in OSS. For example, they might come across a potential target's repositories when searching for OSS they can use, inter-act with, or contribute to, or they might become aware of the firm through activity in forums or mailing lists or when interacting with other community members and word of mouth. As a result, being active in OSS gives them access to addi-tional channels to become aware of potential targets that acquirers not active in OSS lack. Second, being active in OSS allows them to better and faster evaluate a target's OSS activities leading to earlier acquisitions. Acquirers active in OSS themselves have experience in developing OSS and the capabilities to evaluate the quality of the OSS produced by other firms, a capability acquirers not active in OSS most likely lack. Both arguments find support in the interviews, where acqui-rers themselves active in OSS highlighted the role of OSS for sourcing targets and evaluating targets (see Section 3.3.2).

I find no evidence that direct collaboration of the acquirer and the target in OSS would be associated with earlier or delayed acquisitions. When it comes to acqui-sition timing, different potentially counteracting mechanisms could be present: On the one hand, collaboration allows the acquirer to obtain valuable information about a potential target, which can help to overcome information asymmetries in the acquisition process (Hernandez and Shaver, 2019; Zaheer et al., 2010), which, in turn, would lead to earlier acquisitions. On the other hand, building a collaboration takes time, which itself delays a potential acquisition. Given the insignificant result, those mechanisms either equal each other out, or play no role after all. Furthermore, one needs to consider that having collaboration in place might actually render an acquisition unnecessary in some cases; the collaborators already have access to each other's technology, and an acquisition to get access to a firm's technology or resources would be unnecessary as long as no other motives for an acquisition, such as a competitive threat—for example, another potential acquirer showing interest in the firm—or one firm running into finan-cial problems and requiring the collaborating partner to "safe" them through an acquisition appear. This argument finds support in the low number of acquisitions in the sample where collaboration took place before the acquisition—in only 17 out of 549 acquisitions were target and acquirer were active in OSS, collaboration took place before the acquisition. I did also not find collaboration prior to an acquisition particularly relevant for the acquisition process in my interviews.

4.6   Discussion and Conclusion

4.6.1   Summary of Findings

In this chapter I quantitatively examined the influence of OSS activities on the probability and timing of acquisitions utilizing firm and acquisition data from Crunchbase and OSS development data from GitHub. I showed that young firms active in OSS development are associated with a higher acquisition hazard than firms with no activity in OSS development. I also showed that increasing OSS activity is associated with increasing acquisition hazard. While I cannot claim to fully prove causality, the result was highly significant across various robustness checks, including matching and utilizing a competing risk analysis. I discussed five potential mechanisms behind these findings and concluded that only four of them are potential drivers of the result, while I rejected two of them. Higher visibility of firms active in OSS, better possibilities to evaluate them, higher quality of products developed when interacting with the OSS community, and wider user adoption of OSS are all potential mechanisms driving the result. Hype or OSS-companies being sold more often due to financial difficulties caused by not being able to monetize the OSS do not seem to be drivers of the effect.

I do not find evidence that license choice or a firm's ability to attract outside contributions would influence acquisition hazard. I argued that this, together with the first finding, indicates that acquirers care more about a target's OSS development and the quality of the engineers they are about to acquire than a firm's licenses or their interaction with the community.

Lastly, I showed that acquirers themselves active in OSS acquire targets active in OSS earlier than acquirers not active in OSS. I argued that firms themselves active in OSS have better access to targets active in OSS and possess the necessary capabilities to evaluate a target's OSS activities.

4.6.2 Contribution

I suggest that this study makes several contributions and, thereby, adds to a call to produce novel theory at the intersection of open innovation and strategy research (Alexy et al., 2020) by exploring a phenomenon at the intersection of OSS and acquisition research.

First, it contributes to acquisition research. The study expands the understanding of firm characteristics influencing their attractiveness as a target and likelihood of getting acquired (Chakrabarti and Mitchell, 2013; Fischer et al., 2020; Hernandez and Shaver, 2019; Ransbotham and Mitra, 2010; Rogan and Sorenson, 2014). While previous research has identified the role of reducing uncertainty for acquisitions to take place and identified several factors such as patents or regulatory approval to do so (Fischer et al., 2020; Ransbotham and Mitra, 2010), the findings from this study suggest that free revealing of firms in OSS decreases uncertainty around young firms as acquirers can more easily find them and collect information about the firm's qualities, and, thus, increases the likelihood of young firms getting acquired. It further adds to acquisition research by expanding the understanding of the role of an acquirer's capabilities for the timing of acquisitions (Hlavka, 2019). While previous research has found that general technological evaluation capabilities of the acquirer are not significantly related to target age (Hlavka, 2019), which was confirmed in this study with the control variable for sub-industry overlap not being significant, this study provides first evidence that more specific capabilities and activities of the acquirer like the participation in OSS development do so. Results suggest that an acquirer can obtain earlier access to potential targets active in OSS and build distinct capabilities to evaluate other firm's OSS activities by being active in OSS themselves better and faster, which results in earlier acquisitions of firms active in OSS development. Both firms—target and acquirer—being active in open innovation can thus be a factor reducing information asymmetries between targets and acquirers in acquisitions (Chondrakis, 2016), potentially rendering other, more costly ways of getting access to relevant information, such as direct collaboration, unnecessary.

Second, it contributes to OSS, and broader open innovation research, by adding to the call in the open innovation literature for more investigation into the strategic dimensions of a firm's openness decision (Chesbrough, 2006; Dahlander and Gann, 2010). Specifically, it provides first evidence about the role of a firm's openness—in this case participation in OSS development—for its probability of getting acquired. Being active in OSS increases the chances of getting acquired, a key exit strategy for many founders of, and investors in young firms. Moreover, it expands the list of capabilities a firm can build by being active in OSS (Dahlander and Wallin, 2006; Nagle, 2018a) by providing evidence that being active in OSS can help acquirers in their target search and evaluation process.

Furthermore, the study makes a methodological contribution providing input to a call by Alexy et al. (2020) to explore novel data sources to create theory on open innovation. Particularly, my approach to identifying firm-level OSS activities is novel in that regard. Data sources for firm-level data, in particular, data on corporate transactions, usually do not have data on a firm's activities in OSS. Data on OSS activities is usually on an individual level and does not provide information on (characteristics of) corporate transactions. I solve this challenge by combing GitHub OSS development data with Crunchbase firm-level data utilizing email-domain-based matching. Detailed time-stamped information in the combined database allows the observation of individual contributor's activities on a highly granular level and aggregating those activities on firm-level.

4.6.3   Managerial Implications, Limitations, and Outlook

Managerial implications. The findings in this study have several managerial implications. Founders and their investors who consider selling their firm as a promising exit strategy need to be aware of the potential influence of their firm's OSS activities on their likelihood of getting acquired. Participating in OSS development might not only be useful to attract talent, but also to showcase (OSS) capabilities to potential acquirers. This in turn might lead to a higher share of acqui-hires of companies active in OSS development. Acquirers need to be aware that being active in OSS themselves can particularly be relevant when it comes to acquisitions of targets active in OSS development. Being active in OSS themselves may enable them to become aware of other market participants and potential targets faster and allow them to better evaluate a potential target's quality leading to better acquisition decisions and earlier acquisitions.

Limitations. This study is not without limitations. It is limited to the early years of young firms, limiting generalizability across the lifecycle of firms. With the OSS ecosystem growing further, the availability of data should allow applying similar analysis over a longer lifespan of firms. Similarly, the study is limited in that I only explored the GitHub development platform. While GitHub is currently the largest OSS development platform globally, it is yet different from other platforms, particularly platforms for other types of open innovation like crowdsourcing of ideas. This limits generalizability across other platforms and other types of open innovation.

The study also has some methodological limitations. Despite my effort to come close to causal inference by constructing several robustness checks, such as an analysis utilizing matching, utilizing additional data sources, such as Orbis for industry and revenue data, or conducting a competing risk analysis, my analysis does not allow to quantitatively differentiate the effects of different mechanisms, which would all result in the observed main effect. In addition, potentially unobserved variables might influence the effect. For example, founders or managers of firms active in OSS might have a different attitude towards acquisitions than managers of firms not active in OSS. Furthermore, the use of commits to identify firms active in OSS development has—while regularly being used in management and information systems research (e.g., Daniel et al., 2018; Nagle, 2019)—some limitations. Firms might also participate in OSS by being active in mailing lists, forums, or at events and could technically—while unlikely—run an OSS project without actively contributing to the project themselves.

Future research. This study represents the first quantitative evidence about the role of OSS development for acquisitions. However, more research is needed to solidify these initial results and quantitatively understand which of the mechanisms I suggested drive the main effect of firms active in OSS being associated with a higher acquisition likelihood. Furthermore, the relevance of additional characteristics of a firm's involvement in OSS, such as the position of a firm in a network of firms being active in OSS and connected by communities, for a firm's likelihood of getting acquired might be interesting to conduct further research on. Similarly, this study only differentiates between firms actively contributing to OSS and those not contributing. One factor to examine could be the influence of *just using* OSS for internal software development for a firm's probability of getting acquired. Certain benefits of OSS, such as the fast and easy access to pre-built solutions of typically high quality, can already be obtained by integrating OSS into a firm's internal software development and do not require actively engaging in OSS development. Also, the mix of OSS and proprietary software development and the share of each type of software development activity within a firm could be an interesting aspect to look at. Additionally, more research needs to be conducted across the full lifecycle of firms, particularly if acquisition likelihood remains higher for firms active in OSS development across a company's lifecycle and how a target's involvement in OSS influences the timing of acquisitions in the long run. Lastly, the study at hand focuses on the role of OSS before acquisitions. Further research is required to understand the consequences of acquisitions on OSS development. Here, how OSS development activities of acquired firms, acquirers, and the community around them evolve after acquisitions seems particularly relevant.

Mixed-methods Study: The Effect of Acquisitions on Open Source Software Development

5

This chapter focuses on the post-acquisition phase aiming to create an understanding of how acquisitions affect contributions to OSS. The chapter presented hereafter is based on joint work with Joachim Henkel (TUM) and Henning Piezunka (INSEAD).[1]

---

[1] As the first author of this paper, I initially came up with the idea to do research at the intersection of acquisitions and OSS, focusing on both, before and after acquisition events. Also, I designed and conducted all data collection and analysis. The respective chapters (methodology and quantitative and qualitative results) have been written by myself with revisions by my co-authors, while the other chapters (introduction, theoretical background, discussion) are joint work by my co-authors and myself. See signed declaration in digital Appendix G.

Further acknowledgments: We—Joachim Henkel, Henning Piezunka, and I—are grateful to Helge Klapper and Martin Wallin for valuable comments on earlier versions of this paper. We thank Georgios Gousios, founder of the GHTorrent project, for giving us the opportunity to discuss our approach to utilizing GitHub data with him and granting us access to his GHTorrent MongoDB server. We are grateful to Daniel Obermeier for setting up and testing an algorithm to check company affiliation of GitHub users on LinkedIn for the verification of our domain-based user-company affiliation logic (see https://github.com/DanielObermeier/github-profile-scraper). We are grateful to Crunchbase for kindly providing an API key for their database and granting us the right, under the terms of an academic license, to publish aggregate information based on their data.

Electronic supplementary material The online version of this chapter (https://doi.org/10.1007/978-3-658-35084-0_5) contains supplementary material, which is available to authorized users.

M. Vetter, *Acquisitions and Open Source Software Development*, Innovation und Entrepreneurship, https://doi.org/10.1007/978-3-658-35084-0_5

5.1   Introduction

OSS is often developed by a community of individuals who volunteer to contribute without being remunerated and employees of sponsoring firms (Dahlander and Magnusson, 2005; Fisher, 2019; O'Mahony and Bechky, 2008). Sponsoring firms benefit from their employees' engagement as they develop internal and external resources; that is, they build up internal capabilities in developing OSS projects, and help to develop the external OSS project and the community that supports it (Dahlander and Wallin, 2006; Nagle, 2018a). While these resources are of great value to the sponsoring firm, they are often of even greater value to other firms (Singh and Montgomery, 1987) as indicated by the commonality of acquisitions where a firm sponsoring an OSS project is the target.

We examine *how such acquisitions affect contributions to OSS projects.* This question is important on a phenomenological level because an acquisition may affect contributions to the project, a crucial factor for the acquirer, the target, the community, and users relying on the OSS. On a theoretical level it is interesting as we can learn about the evolution of OSS communities, the strategic value for firms to engage in OSS, and the reaction of communities when the control rights of firms that sponsor them change.

Acquisitions may affect contributions to the OSS project both by the acquired sponsoring firm and by the community. Research on acquisitions has examined how the behavior of the acquired target as well as of the stakeholders associated with the target changes (Hernandez and Menon, 2018; Kim, 2019; Valentini, 2016), but this research has yet to cover contributions to OSS. Thus, it remains unclear how contributions to OSS projects from different stakeholders, sponsors, and community members, in an open innovation environment are affected. Research on OSS has focused on the governance and control of OSS projects (He et al., 2020; O'Mahony and Ferraro, 2007; Shah, 2006) and the motivation of contributors (Lakhani and Wolf, 2005; von Krogh, Haefliger, Spaeth, and Wallin, 2012), including the motivation of sponsoring firms and their employees (Bonaccorsi, Giannangeli, and Rossi, 2006; Spaeth et al., 2015), but little is known about the effect of changes in control rights at the sponsoring firm on OSS projects (as it occurs in the case of an acquisition).

Prior research does not allow for theory-based predictions of how acquisitions affect contributions—there are several counteracting effects, and possibly unknown effects in addition. Contributions may go up: the target may increase its contributions, for instance, if the acquirer has a strategic interest in the OSS projects and thus provides the target with additional resources to foster its development (Capron et al., 1998). Also, the community may increase its contributions as developers may appreciate the increased prominence of the project or perceive the acquirer as an attractive employer to whom they wish to signal their skills (Lerner and Tirole, 2002). But contributions may also go down: the target may (be forced to) reduce its contributions if the acquirer sees no value in the specific OSS project, or—in the extreme—sees it as a threat and seeks to "kill" a substitutive project (Cunningham et al., 2020) and thus rather extracts resources from the target and the OSS project. Also, the community may decrease its engagement as it may perceive the acquisition of the sponsoring firm as a disruption to the informal collaboration with the sponsoring firm on which the OSS project was based (Shah and Nagle, 2020; Zhang and Zhu, 2011). Thus, despite the frequency and economic significance of acquisitions of OSS-sponsoring firms,[2] we do not know about the effect of these acquisitions. Anecdotal evidence suggests that it can go both ways: For example, the acquisition of Sun Microsystems, the sponsor of the Open Source MySQL database management system, by Oracle in 2010 led to an outcry in the community, triggering several key employees to resign, and sparked an ongoing dispute about the future openness of MySQL. In contrast, the acquisition of Xamarin by Microsoft in 2016 was perceived positively by the Mono community and employees, some of whom made impressive careers at Microsoft, and resulted in the open sourcing of formerly proprietary parts of Mono after the acquisition.

In the absence of a clear theoretical prediction, we depart from traditional hypothesis testing. Instead, we conduct a rigorous quantitative analysis following the 'just the facts' approach (Oxley et al., 2010) using a Difference-in-Differences analysis. We study the effect of 138 acquisitions of sponsoring firms affecting 347 OSS projects between 2012 and 2017. We match these OSS projects with 1,415 OSS projects not affected by acquisitions. We complement our findings with qualitative data from 52 interviews.

In the first part of our analysis, we find that subsequent to the acquisition of a sponsoring firm (i.e., the target) contributions to the OSS projects—by the acquired sponsoring firm and by the community—decrease on average. The decrease is stronger for contributions from the target; in fact, when controlling for lagged contributions by the acquired target, the association between acquisition and contributions from the community becomes insignificant. This finding suggests—and our interviews support this interpretation—that instead of opposing acquisitions *per se*, community members are rather reacting to the target's post-acquisition

---

[2] The website index.co (https://index.co/market/open-source/acquisitions last accessed 06/15/2020) documents 100 acquisitions of companies with an OSS business model for the years 2012–2018, with deal values of up to $34 billion for the acquisition of Red Hat by IBM 2018 (one of the ten biggest acquisitions by deal value in 2018).

behavior and participation in the OSS project. The general decline of contributions thus seems to be mainly driven by a reduction of contributions by the sponsoring firm following the acquisition.

In the second part of the analysis, we examine the heterogeneity among acquisitions along three dimensions that should affect the outcomes of the acquisition: Organizational integration of the target, which is generally known to affect acquisition outcomes (e.g., Puranam et al., 2006); OSS activity of the acquirer, since it reflects the acquirer's experience in handling OSS projects; and the license type of the target's OSS project—copyleft vs. permissive—since it strongly affects the possibilities to use the focal OSS commercially (e.g., Lerner and Tirole, 2005). We find all three moderators to be relevant: Contributions to OSS projects decrease more after acquisition if the target becomes organizationally integrated, if the acquirer is active in OSS, and if the target's OSS project is under a permissive license.

Our interviews suggest a unifying theme behind these findings. Acquisitions seem to differ by the amount of resources the acquirer extracts from the target, which is driven by differences in acquirers' ability and tendency to do so: Integration sets the organizational framework for resource extraction; OSS experience makes it likely that the acquirer can redeploy the target's resources, code, and employees, to other projects; and a permissive license makes it easier to reuse the project's code commercially. Taken together, our data suggests that it is the extraction of resources by acquirers that shapes the evolution of contributions by the target and the community, while it also allows us to examine and rule out alternative mechanisms.

Our study contributes to research on OSS and beyond. We contribute to research on the survival and success of OSS (Fang and Neufeld, 2009; Ho and Rai, 2017; Shah, 2006) by examining the impact of acquisitions, characteristics of sponsoring firms, and types of licenses. We also contribute to research on the OSS related strategies of firms (Alexy, West, Klapper, and Reitzig, 2017; Fisher, 2019; Gambardella and von Hippel, 2019; Nagle, 2018a, 2018b; West, 2003) as we identify two types of strategies that acquirers pursue: resource extraction vs. continuation of the target's activities. We also contribute to research on the management of external resources (Dahlander and Magnusson, 2005; Dahlander and Wallin, 2006; Franke, Keinz, and Klausberger, 2013; Klapper and Reitzig, 2018; West and O'Mahony, 2008) as we illustrate the transferability of external resources, i.e., that acquirers can maintain a community to which they did initially not contribute. Finally, we contribute to the debate on the level of integration and its effects on targets (Datta and Grant, 1990; Graebner, 2004; Puranam et al.,

2006) as we show that tighter integration of the target by the acquirer can reduce the availability of external resources.

5.2 Theoretical Background

5.2.1 Motivation of Contributors to OSS

To lay the foundations for our study of how acquisitions may affect OSS contributions, we review the literature on motivations to contribute to OSS development. The evolution of OSS projects depends on contributors' tendency to join and to continue contributing (Ho and Rai, 2017; Seo, Nagle, and Shah, 2020; Oh and Jeon, 2007; Shah, 2006). In line with prior research, we distinguish two types of contributors: sponsoring firms and community members.

Sponsoring firms dispatch their developers to contribute code to the projects or let them take over managerial and administrative tasks in the community. By doing so, sponsoring firms can develop internal and external resources; that is, they build up internal capabilities in developing OSS projects, and help to develop the OSS project and the community that supports it (Dahlander and Wallin, 2006). The social capital the sponsoring firm accumulates when forming a relationship with the community can constitute an external resource for the firm (Alexy et al., 2017; Dahlander and Wallin, 2006; Fisher, 2019). Together, such resources have shown to translate into strategic advantages as firms engaging in OSS development are able to attract valuable talent (Ågerfalk and Fitzgerald, 2008), cut product development or sourcing cost (Dahlander and Magnusson, 2005; Gambardella and von Hippel, 2019), increase absorptive capacity and productivity (Nagle, 2018a, 2018b), gain faster user adoption (West, 2003), or increase demand for complementary services and products (Andersen-Gott et al., 2012). These strategic advantages sponsoring firms can build from their engagement in OSS can translate into potential motives to acquire them.

Beyond the sponsoring firms, often individual community members help to develop the OSS project. Research has identified a variety of reasons why people begin and continue to contribute (e.g., Bateman, Gray, and Butler, 2011; Lakhani and Wolf, 2005; Ren, Kraut, and Kiesler, 2007). Typically, people use the OSS code themselves and then begin contributing when they see potential to improve it (Kane and Ransbotham, 2016). They continue to contribute because they become attached to the community and bond with their peers (Bateman et al., 2011; Ren et al., 2012; Ren et al., 2007; Shriver et al., 2013; Zhang and Zhu, 2011). Further important motives are the intellectual stimulus of coding and the improvement of programming skills (Lakhani and Wolf, 2005). Additionally, signaling their capabilities to potential employers and gaining status and reputation in the community can be an important reason to work on OSS projects (Gallus, 2017; Lerner and Tirole, 2002). Given the important role of individuals in developing OSS software, it is crucial to understand how they behave following an acquisition.

While the willingness of sponsoring firms and of community members to contribute provides a strong basis for the survival and success of OSS projects, research also shows that the willingness to contribute is not necessarily robust, rendering OSS projects fragile. A particular source of fragility is that contributors are in part motivated by the presence of other contributors and the overall project activity (Oh and Jeon, 2007; Zhang and Zhu, 2011). Thus, the departure of key contributors can result in a chain reaction with more and more contributors leaving (Oh and Jeon, 2007), endangering the very existence of the OSS project. As a result, if a sponsoring firm ends or reduces its engagement in an OSS project, the project can be critically endangered as a substantial part of the contributions vanishes. Such dramatic changes in the sponsoring firm's level of contributions to the project might be triggered by an acquisition where the acquirer decides against continuing the support of the project or extracts significant resources from it. Yet, if the acquirer commits more resources to an acquired target's projects, a project might flourish after an acquisition and attract further contributions from the community. The community might also increase its activities if they perceive the acquirer as an attractive employer to whom they wish to signal their skills (Lerner and Tirole, 2002). Hence, while informative, prior research on OSS does not allow clear predictions about how the contribution behavior of firms and individuals and their interaction will evolve after acquisitions.

5.2.2   Consequences of Acquisitions

To inform our understanding of how an acquisition may affect an OSS project we review the literature on the consequences of acquisitions for acquisition targets. Acquisitions can result in considerable changes at the target: The organizational structure of the target may be modified (Karim, 2006), employees may leave the firm (Ernst and Vitt, 2000), the type of innovation pursued by the target may change (Choi and McNamara, 2018), and its performance may change (Haspeslagh and Jemison, 1991; King et al., 2004; Véry, Lubatkin, Calori, and Veiga, 1997). The acquisition does not just result in changes at the target itself, but also affects its stakeholders (Bettinazzi and Zollo, 2017; Hernandez and Menon,

2018; Rogan and Greve, 2015; Valentini, 2016). Recent research shows that customers' tendency to collaborate with the target and its acquirer changes dramatically (Rogan and Greve, 2015), and that competitors of the target or the acquirer may react by changing their innovation strategy (Valentini, 2016).

The degree of change triggered by an acquisition depends on various contingencies. Studying the effect of structural integration, prior research identified a negative impact on post-acquisition innovation (Paruchuri et al., 2006; Puranam et al., 2006). The technological relatedness of target and acquirer has a nonlinear impact on post-acquisition innovation, where acquisitions of targets with related technologies result in higher levels of post-acquisition innovation compared to acquisitions of targets with a very similar or unrelated technology base (Ahuja and Katila, 2001). Furthermore, the acquirer's ability and experience to manage the acquisition process can lead to better outcomes for post-acquisition innovation (Haleblian and Finkelstein, 1999; Zollo and Singh, 2004). More recently, Bettinazzi and Zollo (2017) found that stakeholder orientation of the acquirer can improve post acquisition performance. Lastly, key contingencies are the acquirers' intentions—in particular whether to redeploy or divest resources following an acquisition (Capron et al., 1998; Capron and Mitchell, 2001; Karim and Capron, 2016). Yet, intentions are often not observable, and the acquirer may not succeed in realizing them. Overall, while highly informative, the research on acquisitions does neither provide clear guidance on how the target and the community (who is a main stakeholder) may respond to acquisitions nor how the acquirer may treat the acquired target's OSS projects.

Summarizing, the literature on OSS shows that sponsoring firms and their actions are important for the ongoing success of the OSS projects they are involved in. While typically a source of strength, having a sponsoring firm also renders OSS projects exposed to events affecting the sponsoring firm. The research on acquisitions shows that acquisitions often affect the target and its stakeholders significantly, with the effect being contingent on how control rights are exerted. Yet, the literature does not allow us to conclude how acquisitions influence the target's or the community's contributions to OSS projects after an acquisition.

5.3   Data and Method

To investigate the impact of acquisitions on contributions to OSS projects, we collect data on OSS projects that were sponsored by firms that were acquired. We then use Difference-in-Differences (DiD) regressions to compare changes in contributions to these OSS projects to changes in contributions to OSS projects in a control group.

5.3.1   Sample Construction

For data on acquiring and sponsoring firms as well as on acquisitions we rely on Crunchbase, for data on OSS activities on GitHub. GitHub is the most widely used platform for public OSS development, for individuals as for firms of various sizes (Burton et al., 2017).[3] GitHub data allows us to track each contributor's activities over time and to associate individual contributors with firms by matching contributors' email addresses with companies from Crunchbase based on their domain name, and thus to identify companies sponsoring OSS projects. We collected GitHub data[4] for the years 2011 to 2018 (the GitHub Archive is available as of 2011), and acquisition data for the years 2012 to 2017 to observe one year of OSS activity prior to and after each acquisition.

Our unit of analysis is the individual OSS project on GitHub, defined by the repository containing all project files. In order to identify projects affected by an acquisition, we first identify firms that were acquired and that had been actively contributing on GitHub in the 12 months prior to acquisition. We measure contributions by number of code commits (c.f. Daniel et al., 2018; Nagle, 2019).[5] We match email domains of targets obtained from Crunchbase with email addresses of GitHub users, an approach used by Nagle (2018a) for the case of Linux.[6] We

---

[3] GitHub hosts over 100 million software projects (so called repositories) and is the world's largest OSS development platform. GitHub provides a rich empirical setting for management research and as a result has recently sparked increased interest among researchers in the field of management (e.g., He et al., 2020; Nagle, 2019).

[4] To create a rich set of information about the OSS activities on GitHub we use data from three different data sources containing GitHub data, namely the GHTorrent project, GHArchive and the GitHub API. Our main source is the GHTorrent project (http://ghtorrent.org/) (Gousios, 2013). GitHub Archive (https://www.gharchive.org/) data is used to obtain email-domains used by GitHub users when committing code on GitHub, while we use the GitHub API (https://developer.github.com/) to obtain license information about the projects (i.e., repositories) hosted on GitHub. All data sources can be matched either on unique commit-"shas" or unique repository names.

[5] A commit is a change in the source code of the project. In GitHub a commit is always linked to the person who wrote the code, be a user with direct write access to the project or external contributors who have to submit their proposed commits to a project via a pull request.

[6] Some GitHub users do not use their firm email address. To ensure correct labelling also of those users we employ an algorithm that checks company affiliation by matching names of GitHub users to LinkedIn profiles. We checked over 100,000 GitHub users working on projects in our sample and identified ~ 20,000 users working for a sponsoring company in a project without using a company email address.

exclude targets that do not allow us to clearly identify their employees.[7] Next, similar to He et al. (2020), we restrict the sample to independent projects[8] (i.e., not forked from a parent project) and those with at least five members from the target and five not employed by the target.[9] We furthermore require that the target's share of all contributions is above 5% and below 95%.[10] Additionally, as we use the license type for matching, we exclude projects for which we cannot obtain license information (58 projects). Overall, this leaves us with 444 projects linked to 168 acquisitions.

To construct a control group, we create a set of OSS projects with non-acquired sponsors, by applying the same selection criteria as for the treatment group and excluding projects where also a target was active in. We then employ coarsened exact matching (CEM) to create samples from the treated and the control group that are similar with respect to the joint distribution of (potentially confounding) variables on project and firm level (Iacus, King, and Porro, 2012). Specifically, we use five pre-treatment firm and project characteristics for matching: The size of the focal firm, its share of commits to the project, the project's license, the level of commit activity in the year prior to the acquisition, and the commit activity

---

[7] The reasons for not being able to identify employees clearly are as follows: We could not include email providers such as yahoo (acquired by Verizon in 2017) in our dataset, as most commits under the "yahoo.com"-domain are from users of the yahoo email service, not employees. For a similar reason we excluded acquisitions of online-education facilities such as the Flatiron School (acquired by WeWork), as they often use GitHub as a host for their students' projects, as well as companies using a LinkedIn or Facebook page as their homepage or a government or a university domain on Crunchbase. We also excluded firms which used their parent organization's domain (prior owner in case of a carve-out or acquirer) on Crunchbase, as we cannot differentiate between those employees affected by an acquisition and those not being part of the acquired entity.

[8] We exclude forks since they inherit all contributions from their respective parent project and hence create a large number of duplicates.

[9] According to Kalliamvakou, Gousios, Blincoe, Singer, German, and Damian (2016) more than two thirds (71.6%) of GitHub projects have only a single participant (its owner). By filtering we ensure that the firm has invested considerable resources into a project and avoid selecting projects where employees contribute as hobbyists outside of their work. Furthermore, we ensure that the projects were successful in attracting contributions from the community, which allows us to capture the effect of an acquisition on contributions from the acquired target as well as on those from the community.

[10] Our results hold for different specifications of our sample, particularly changes in the coarsening of our matching parameters, including smaller projects or focusing only on larger projects.

growth trend of the project.[11] Using these coarsened variables to match on, we find one or more matches for 347 treated projects, associated with 138 acquired targets. Our matched control group is composed of 1,415 projects associated with 695 non-acquired firms. We employ CEM weights to account for varying strata size (Blackwell, Iacus, and Porro, 2009).

5.3.2   Variables

Dependent Variable: Contributions to projects. To measure *Contributions,* we capture the monthly number of commits to a project, both in total and separately for commits made by employees of the focal firm and those made by other contributors. We use logarithms as those variables have a right-skewed distribution.

Independent Variables: The main independent variables we use are *Treated,* which is used to differentiate the treatment and the control group, and *PostAcquisition,* which is used to differentiate the months prior and after the acquisition. Their interaction—the DiD-term—captures the effect of the acquisition, namely the change in contributions by the treatment group compared to the control group after the acquisition. The variable *Time* (in months) controls for time trends. We estimate our models with project-level fixed effects.

Moderators. In the second part of our analysis, we evaluate three moderating variables: *Prior OSS activity of the acquirer* is operationalized as the level of commit activity on non-private GitHub repositories by the acquirer in the year

---

[11] We measure the size of the focal firm, a variable commonly used in post-acquisition innovation research (e.g., Bertrand, 2009; Kapoor and Lim, 2007), by its number of employees using five coarsened size categories. The focal firm's share of commits in the year prior to the acquisition captures its relative importance for the project. We coarsen this variable to four categories. The project's license type affects the acquirer's options to include the code into its proprietary software. Using the Wikipedia categorization of OSS licenses, we differentiate between permissive (e.g., MIT, Apache, BSD), copyleft (e.g., GPL, AGPL, LGPL), and mixed license projects. We obtain licenses via the GitHub API and collect additional data from GitHub repositories and by searching the projects' commit history (in particular for defunct projects). Lastly, we need to ensure that treatment and control projects are similar with respect to the level and the trend of their commit activity prior to the acquisition. We do so by capturing the overall level of activity and the trend by matching on the coarsened number of commits in the two half-years prior to the acquisition (for each half-year eight categories to match on after coarsening).

prior to the acquisition relative to its size as measured by the number of employ-ees.[12] In order to create a dummy for two distinct groups—OSS active acquirer and non-active acquirer—we consider all firms with a score of 1 or higher as OSS active (119 projects linked to 41 acquisitions) and all others as non-active (190 projects linked to 93 acquisitions).[13] For robustness checks we also use the absolute number of commits as a measure for the level of OSS activity of the acquirer.

License type is operationalized in line with the license types we already used in the matching (permissive and copyleft). Due to the small number of only six observations, we drop mixed-license projects.

Structural integration is a binary variable that is 1 if the target was structurally integrated into the acquirer and 0 otherwise. We gathered the data from press releases, newspapers, and professional industry publications (Bettinazzi and Zollo, 2017; Paruchuri et al., 2006; Puranam et al., 2009).[14]

5.3.3 Quantitative Analysis

To estimate the effects of acquisitions on post-acquisition contributions to OSS projects, we adopted a DiD approach commonly used in acquisition research (Bertrand, 2009; Colombo and Rabbiosi, 2014). We observe for treated and non-treated OSS projects the number of commits for 12 months before and after the

---

[12] As Crunchbase reports size categories of firms on a quasi-logarithmic scale we create similar categories for the acquirer's OSS activity to set them in relation. For instance, if Crunchbase reports a company size of 51–100 employees and this company made 51–100 commits in the year prior to the acquisition we report a level of OSS activity of 1. If the number of commits was one category below (21–50), we report a level of OSS activity of 0; if it was one category above (101–250), we report a level of OSS activity of 2, etc.

[13] For four acquirers, Crunchbase did not provide size information. We dropped the eight projects linked to these acquirers and their matched control group projects in regressions where we use the level of the acquirer's OSS activity relative to the acquirer's size as moderator.

[14] If the press release reported a statement such as "Red Hat plans to make Eclipse Che and the Codenvy enhancements central to its tooling strategy and extend and integrate the workspace management technology across tools and platforms", then we concluded that a structural integration had occurred. However, if the press release included statements such as "Pentaho will retain its existing brand and continue to operate independently… under the leadership of its CEO", then we recorded this acquisition as structural separation. As a further validity check, we examined the website of the target to verify if the target was indeed integrated (e.g., if the target's website links to the acquirer's website, was shut down, or clearly indicates the target now belongs to the acquirer).

acquisition.[15] We estimate clustered standard errors and cluster on focal firm-level (i.e., the target or the non-acquired firm from the control group) using the Stata package 'reghdfe' (Correia, 2017). To test the influence of the moderators, we perform regressions with separate DiD interaction terms for each of the two values of the respective moderator.[16]

5.4   Quantitative Results

5.4.1   Descriptive Results

Tables 5.1 and 5.2 provide descriptive statistics and correlation coefficients, separately for the treatment and the control group. The correlations are similar for both groups: t-tests to compare correlation coefficients for the pre-treatment contributions to projects across the treatment and the control group as well as a Hotelling test to jointly compare the variables across treatment and control group are insignificant (p-value$_{\text{t-test LnCommits}}$ = 0.629; p-value$_{\text{t-test LnFocalCommits}}$ = 0.594; p-value$_{\text{t-test LnOtherCommits}}$ = 0.714; p-value$_{\text{Hotelling}}$ = 0.776).

---

[15] In line with McCarthy and Aalbers (2016), who conduct research on patent-based post acquisition innovation performance, we focus only on the contributions to OSS projects in the first year after an acquisition. We do so for three reasons: (1) given the high prices for technology-focused firms, acquirers want to show results in the first year after the acquisition to signal the success of the acquisition (McCarthy and Aalbers, 2016; Valentini, 2012). As a result, if there is an effect, it should be observable within the first year after an acquisition; (2) increasing levels of employee turnover due to differences in culture and internal political and hierarchical relations, mean that any benefits of an acquisition are either directly reaped, or lost (McCarthy and Aalbers, 2016); (3) the further we move from the acquisition the greater the number of potentially unidentifiable variables influencing the outcomes of the acquisitions (McCarthy and Aalbers, 2016). The last argument is particularly relevant when conducting research on OSS projects, as the high project development dynamics taking place in OSS with projects regularly losing traction also in the absence of acquisitions (e.g., Oh and Jeon, 2007) might further increase the number of unidentifiable variables the further we move from the acquisition. Overall, we focus on the first year, because we expect an observable effect from the acquisition in the first year.

[16] For the analysis of moderators, we match some projects of the control group with projects in both parts of the treatment group (e.g., with a project where the acquirer is OSS-active and one where the acquirer is not OSS-active). The overall number of observations in the regression for the control group is thus 1,966 projects from the control group instead of our 1,415 individual projects. As CEM-weights are adjusted accordingly regression results are not influenced.

Table 5.1 Descriptive statistics & correlation coefficients—Treated

Variables		Descriptive statistics					Correlation coefficients							
		N_projects	Mean	S.D.	Min	Max	[1]	[2]	[3]	[4]	[5]	[6]	[7]	[8]
[1]	Monthly contributions to project (ln)	347	2.38	1.83	0	7.46	1.00							
[2]	Contributions by focal firm (ln)	347	1.80	1.71	0	7.33	0.88 (0.00)	1.00						
[3]	Contributions by community (ln)	347	1.59	1.62	0	6.91	0.84 (0.00)	0.58 (0.00)	1.00					
[4]	Time (month)	347	0.50	6.92	−11	12	−0.24 (0.00)	−0.23 (0.00)	−0.17 (0.00)	1.00				
[5]	PostAcquisition	347	0.50	0.50	0	1	−0.23 (0.00)	−0.22 (0.00)	−0.16 (0.00)	0.87 (0.00)	1.00			
[6]	Integrated (1 = integrated)	347	0.64	0.48	0	1	−0.03 (0.02)	−0.01 (0.56)	−0.06 (0.00)	0.00 (1.00)	0.00 (1.00)	1.00		
[7]	OSS-active acquirer (1 = active)	339	0.35	0.48	0	1	−0.00 (0.94)	0.00 (0.81)	−0.01 (0.22)	0.00 (1.00)	0.00 (1.00)	0.44 (0.00)	1.00	
[8]	License (0.5 = mixed license; 1 = permissive)	347	0.78	0.41	0	1	−0.05 (0.00)	−0.06 (0.00)	−0.03 (0.20)	0.00 (1.00)	0.00 (1.00)	0.19 (0.00)	0.07 (0.00)	1.00

Note: N = 347 projects with 8,328 observations. Observations, where the acquirer's size was not available, dropped from analysis in row 7 (N = 8). p-values in parentheses.

Table 5.2 Descriptive statistics & correlation coefficients—Control

Variables		Descriptive statistics					Correlation coefficients					
		$N_{projects}$	Mean	S.D.	Min	Max	[1]	[2]	[3]	[4]	[5]	[6]
[1]	Monthly contributions to project (ln)	1415	2.53	1.85	0	8.32	1.00					
[2]	Contributions by focal firm (ln)	1415	1.96	1.75	0	7.88	0.89 (0.00)	1.00				
[3]	Contributions by community (ln)	1415	1.62	1.64	0	7.55	0.82 (0.00)	0.56 (0.00)	1.00			
[4]	Time (month)	1415	0.50	6.92	−11	12	−0.17 (0.00)	−0.16 (0.00)	−0.12 (0.00)	1.00		
[5]	PostAcquisition	1415	0.50	0.50	0	1	−0.16 (0.00)	−0.15 (0.00)	−0.11 (0.00)	0.86 (0.00)	1.00	
[6]	License (0.5 = mixed license; 1 = permissive)	1415	0.78	0.41	0	1	0.01 (0.00)	−0.00 (0.83)	0.00 (0.73)	0.00 (1.00)	0.00 (1.00)	1.00

Note: N = 1,415 projects with 33,960 observations. p-values in parentheses. For the analysis of moderators (Table 5.4–5.6), we match some projects of the control group with projects in both parts of the treatment group (e.g., with a project where the acquirer is OSS-active and one where the acquirer is not OSS-active). The overall number of observations for the control group is thus 47,184 in those analyses. As CEM-weights are adjusted accordingly, correlation coefficients are not influenced.

5.4.2 Contribution Activity after Acquisitions—Main Effect

We first examine how contributions to OSS projects change after acquisitions. Table 5.3 reports DiD regressions for overall contributions, contributions from the sponsoring firm, and those from the community.

The coefficient of the interaction term, *Treated* × *PostAcquisition*, in Model 1 is −0.271 (p = 0.001), indicating that contributions to OSS projects decline significantly after an acquisition compared to the control group; this corresponds to an average reduction of contributions of 23.5 percent. Figure 5.1 provides an illustration. The curves for the treatment and control group are aligned before the acquisition and diverging afterward.[17]

When focusing exclusively on contributions by the target (Model 2), the coefficient of the interaction term is −0.240 (p = 0.001). In contrast, the relative decrease in the number of contributions by the community after an acquisition (Model 3) is considerably smaller: the interaction term carries a coefficient of −0.133 (p = 0.126), suggesting that the community's contribution activity does not change as much after an acquisition as the focal firm's.

As contributors to OSS projects are known to react to changes in other contributors' contribution levels (Shah, 2006; Zhang and Zhu, 2011), we control in Models 4 and 5, analyzing contributions by the focal firm and the community, respectively, for the time-lagged contributions of the respective other group.

In Model 4, we add the 1-month-lagged contributions by the community to the specification of Model 2 to test how far the contribution behavior of the community influences—or in any case, allows to predict—the contribution behavior of the sponsoring firm. We find that while prior contributions by the community partly predict contributions by the sponsor, the interaction term *Treated* × *PostAcquisition* still carries a negative and significant coefficient of −0.186 (p = 0.025), suggesting that the drop in contributions by the focal firm after an acquisition cannot fully be predicted by a preceding drop in contributions from the community; rather, the drop in contributions from the focal firm appears to be linked to the acquisition event itself.

---

[17] Note: 90% confidence interval error bars. Month −1 are the last 30 days prior to the acquisition, month 1 are the first 30 days starting with the acquisition.

The graph also shows that the time trend for commits to projects in our sample, both treatment and control projects, is negative over the 24 month period, consistent with the negative and significant coefficient in all models in Table 5.13. We explain this negative trend over time with the life cycle of OSS projects, many of which lose traction over time. The (nonlinear) effect of the overall reduction in contributions over time is also captured by the term *PostAcquisition,* which is negative in all models.

Table 5.3 Number of contributions to a project before and after acquisition: Main regressions

Variables	(1) Total contrib. (ln)	(2) Contributions by focal firm (ln)	(3) Contributions by community (ln)	(4) Contributions by focal firm (ln)	(5) Contributions by community (ln)
Time	−0.039*** (0.005)	−0.034*** (0.005)	−0.023*** (0.004)	−0.027*** (0.004)	−0.014*** (0.003)
PostAcquisition	−0.114* (0.068)	−0.124* (0.062)	−0.118** (0.055)	−0.079 (0.055)	−0.064 (0.048)
Treated × PostAcquisition	−0.271*** (0.104)	−0.240** (0.095)	−0.133 (0.087)	−0.186** (0.083)	−0.061 (0.072)
Contributions by community (ln) lagged 1 month	–	–	–	0.373*** (0.023)	–
Contributions by firm (ln) lagged 1 month	–	–	–	–	0.322*** (0.018)
Constant	2.600*** (0.027)	2.023*** (0.025)	1.704*** (0.022)	1.389*** (0.038)	1.039*** (0.039)
Project fixed effects	Yes	Yes	yes	yes	yes
Observations	55,512	55,512	55,512	55,512	55,512
R^2	0.569	0.565	0.580	0.619	0.635
Adj. R^2	0.550	0.546	0.562	0.603	0.618
F-statistic	71.27***	75.37***	42.91***	132.32***	103.38***

Note: The variable "treated" is not used as a covariate since it is captured in the project-level fixed effects. Cluster robust standard errors in parentheses. * $p<0.10$, ** $p<0.05$, *** $p<0.01$.

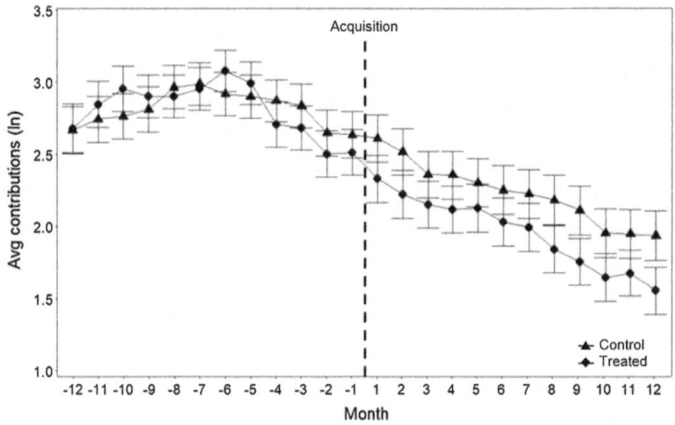

Figure 5.1   Average monthly commits to project around acquisition date

In Model 5, we add, vice versa, the 1-month-lagged contributions by the sponsoring firm as an explanatory variable for contributions by the community. As a result, the coefficient of the interaction term becomes -0.061 ($p = 0.398$), a reduction in absolute size compared to Model 3 by 0.072. The insignificant coefficient of the interaction term in Model 5 suggests that the acquisition event itself has only minor—if any—influence on contributions from the community. Our results are robust to changes in the time lag.

Summarizing, the above results offer quantitative evidence that, on average, contributions to OSS projects go down significantly after acquisition of the sponsoring firm. This decrease is mainly driven by a reduction in contributions by the focal firm. The reduction in contributions by the community is smaller (and insignificant) and appears to be largely a reaction to changes in the focal firm's contributions post acquisition.

5.4.3   Variation in the Change in Contribution Activity after Acquisitions—Moderating Effects

Since we find considerable heterogeneity in the outcomes of acquisitions—58% of projects exhibit a decrease in the number of contributions, while 42% show an increase compared to the control group—we examine this heterogeneity in the second part of analysis in order to unpack the apparent negative effect of acquisitions on contributions to OSS projects. We analyze three moderators that can be expected to affect acquisition outcomes: Structural integration of the target into the acquirer's organization; the acquirer's familiarity with OSS development, measured by its contribution activity; and the license type of the OSS project. Table 5.4 shows the results. The relevant terms are the three-way interactions between *Treated*, *PostAcquisition*, and the respective moderating dummy variable.

For structural integration (Model 1), we find that the decline in contributions is large for projects where the target was integrated ($\beta = -0.393$, p = 0.000, compared to $\beta = -0.271$ in Model 1, Table 5.3), and essentially zero without integration ($\beta = -0.058$, p = 0.715). A Wald test shows that the two coefficients differ significantly (p = 0.062). We find qualitatively the same results when analyzing the two types of acquisitions separately (Tables 5.5 and 5.6): With integration, the DiD estimator is negative and significant for both target contributions ($\beta = -0.359$, p = 0.001) and—in contrast to Table 5.3—community contributions ($\beta = -0.194$, p = 0.029), while it is close to zero in both cases without integration ($\beta = -0.031$, p = 0.815 and $\beta = -0.026$, p = 0.868, respectively).

Model 2 illustrates how prior OSS activity of the acquirer moderates the change in contributions to OSS projects after an acquisition. Surprisingly, we find that projects with an OSS-active acquirer experience a stronger decrease ($\beta = -0.534$, p = 0.000) than projects where the acquirer does not have substantial OSS activity ($\beta = -0.135$, p = 0.264). The coefficients differ significantly (p = 0.026). Again, we find qualitatively the same results when analyzing the two types of contributions separately (Tables 5.5 and 5.6): For acquirers with substantial OSS activity, contributions from the target and from the community are strongly declining after acquisitions ($\beta = -0.399$, p = 0.012 and $\beta = -0.322$, p = 0.007), while for acquirers without substantial OSS activity there is only a minor decline for both ($\beta = -0.140$, p = 0.185 and $\beta = -0.053$, p = 0.617, respectively). Our results are robust to alternative specifications.

Model 3 addresses the type of license as a moderator. We find that projects that use a permissive license are more strongly negatively affected ($\beta = -0.318$, p = 0.002) than projects with a copyleft license ($\beta = -0.058$, p = 0.778), though the coefficients are not significantly different from each other (p = 0.209).[18] We again find qualitatively the same results when analyzing the two types of contributions separately (Tables 5.5 and 5.6): For projects with a permissive license, the DiD estimator is larger in absolute terms both for contributions by the target and for

---

[18] Even though the Wald test is insignificant, our results indicate that the distribution of impact of acquisitions on contributions to OSS projects are different for copyleft and permissive licensed projects. Furthermore, the standard errors from the copyleft licensed projects are large due to the low number of projects in this category ($n_{target} = 75$).

Table 5.4 Number of contributions to a project before and after acquisition: Moderators

DV: Overall contributions (ln) Variables	(1) Moderator: Integration	(2) Moderator: AcquirerOSS (relative to its size)	(3) Moderator: License	(4) Robustness check for Moderator: AcquirerOSS (absolute nr. of commits)
Time	-0.039*** (0.005)	-0.039*** (0.005)	-0.038*** (0.005)	-0.039*** (0.005)
PostAcquisition	-0.114* (0.068)	-0.113 (0.069)	-0.118* (0.068)	-0.114* (0.068)
Treated × PostAcquisition ×	Integrated ($n_{target}=221$) -0.393*** (0.112)	Acquirer active (rel) ($n_{target}=119$) -0.534*** (0.153)	Permissive ($n_{target}=266$) -0.318*** (0.104)	Acquirer active (abs) ($n_{target}=175$) -0.401*** (0.134)
	Not integrated ($n_{target}=126$) -0.058 (0.151)	Acquirer not active (rel) ($n_{target}=220$) -0.135 (0.131)	Copyleft ($n_{target}=75$) -0.058 (0.205)	Acquirer not active (abs) ($n_{target}=172$) -0.139 (0.130)
Constant	2.601*** (0.026)	2.587*** (0.026)	2.582*** (0.026)	2.601*** (0.026)
Project fixed effects	yes	yes	yes	yes
Nr observations	55,512	52,728	55,080	55,512
R2	0.570	0.566	0.562	0.570
Adj. R2	0.551	0.547	0.542	0.551
F-statistic	55.51***	56.73***	56.95***	56.14***

Note: In Model 2, we dropped observations (and the corresponding control group observations) where we did not find size information for the acquirer. In Model 3, we dropped projects with mixed licenses. Cluster robust standard errors in parentheses.
* $p<0.10$, ** $p<0.05$, *** $p<0.01$.

Table 5.5 Number of contributions from the focal firm to a project before and after acquisition: Moderators

DV: Focal firm contributions (ln) Variables	(1) Moderator: Integration	(2) Moderator: AcquirerOSS	(3) Moderator: License
Time	−0.034*** (0.005)	−0.034*** (0.005)	−0.033*** (0.005)
PostAcquisition	−0.124** (0.062)	−0.126** (0.063)	−0.131** (0.063)
Treated × PostAcquisition ×	*Integrated* (n_{target}= 221)	*Acquirer active (rel)* (n_{target}= 119)	*Permissive* (n_{target}= 266)
	−0.359*** (0.106)	−0.399*** (0.159)	−0.307*** (0.094)
	Not integrated (n_{target}= 126)	*Acquirer not active (rel)* (n_{target}= 220)	*Copyleft* (n_{target}= 75)
	−0.031 (0.133)	−0.140 (0.105)	−0.018 (0.189)
Constant	2.023*** (0.024)	2.010*** (0.025)	2.002*** (0.025)
Project fixed effects	yes	yes	yes
Nr observations	55,512	52,728	55,080
R2	0.566	0.564	0.558
Adj. R2	0.547	0.545	0.539
F-statistic	61.73***	55.80***	62.35***

Note: In Model 2, we dropped observations (and the corresponding control group observations) where we did not find size information for the acquirer. In Model 3, we dropped projects with mixed licenses. Cluster robust standard errors in parentheses. * $p < 0.10$, ** $p < 0.05$, *** $p < 0.01$.

Table 5.6 Number of contributions from the community to a project before and after acquisition: Moderators

DV: Community contributions (ln) Variables	(1) Moderator: Integration	(2) Moderator: AcquirerOSS	(3) Moderator: License
Time	-0.023^{***} (0.004)	-0.022^{***} (0.004)	-0.022^{***} (0.004)
PostAcquisition	-0.118^{**} (0.055)	-0.117^{*} (0.056)	-0.121^{**} (0.055)
Treated × PostAcquisition ×	*Integrated* ($n_{target}=221$) -0.194^{**} (0.089)	*Acquirer active (rel)* ($n_{target}=119$) -0.322^{***} (0.119)	*Permissive* ($n_{target}=266$) -0.143^{*} (0.085)
	Not integrated ($n_{target}=126$) -0.026 (0.154)	*Acquirer not active (rel)* ($n_{target}=220$) -0.053 (0.105)	*Copyleft* ($n_{target}=75$) -0.013 (0.178)
Constant	1.704^{***} (0.021)	1.700^{***} (0.021)	1.681^{***} (0.021)
Project fixed effects	yes	yes	yes
Nr observations	55,512	52,728	55,080
R2	0.580	0.578	0.569
Adj. R2	0.562	0.560	0.550
F-statistic	36.34^{***}	35.38^{***}	33.57^{***}

Note: In Model 2, we dropped observations (and the corresponding control group observations) where we did not find size information for the acquirer. In Model 3, we dropped projects with mixed licenses. Cluster robust standard errors in parentheses. $*$ $p < 0.10$, $**$ $p < 0.05$, $***$ $p < 0.01$.

those by the community ($\beta = -0.307$, p $= 0.001$ and $\beta = -0.143$, p $= 0.092$, respectively) than for projects with copyleft licenses ($\beta = -0.018$, p $= 0.925$ and $\beta = -0.013$, p $= 0.943$, respectively).

As in Models 4 and 5 in Table 5.3, we ran the regressions in Tables 5.4, 5.5, and 5.6 with the lagged number of contributions by the community and the sponsor, respectively, as an additional explanatory variable (not reported). In all cases, the DiD term remains significant when the sponsor's contributions are the dependent variable and become insignificant when community contributions are the dependent variable. This finding lends further support to our interpretation that the reduction in contributions from the community can largely be explained as a reaction to a reduction in the sponsor's contributions, but not vice versa.

5.4.4 Robustness Checks

Our results are robust to alternative specifications: They are robust to variations in the coarsening of our variables in the matching, dropping fixed effects in our regression models, and to the inclusion of higher orders of *Time* in the time trend in the regression (not reported). For the role of the prior activity of the acquirer in OSS development, we conduct three robustness checks: First, we use an alternative measure for the OSS activity of the acquirer (see Table 5.4 above, Model 4). Instead of our initial measure where we calculated the activity of the acquirer relative to the company size of the acquirer by its employees, we now use the absolute number of OSS contributions in the year prior to the acquisition and create our dummy by splitting the sample in the middle. The estimators for the regression coefficients are directionally similar to our initial result; -0.401 (p $= 0.003$) for acquirers with substantial activity in OSS prior to the acquisition and -0.140 (p $= 0.284$) for acquirers without substantial activity, the latter one again being insignificant. Furthermore, we conduct a regression with a triple interaction, where we interact our measure of OSS activity relative to the size of the acquirer with the interaction term *PostAcquisition × Treated* (Table 5.7). This triple interaction estimates the change in the impact of an acquisition on OSS contributions when the OSS activity of the acquirer prior to the acquisition changes. The triple interaction estimator is -0.040 (p $= 0.104$; Model 1), indicating on a more granular level than our regression with a dummy that an increasing level of OSS activity of the acquirer leads to a stronger negative impact on OSS projects. We obtain similar results (not reported) for a diff-in-diff-in-diff regression with the absolute level of OSS activity of the acquirer as a third interaction term or the log-transform number of contributions by the acquirer (Model 2: $\beta = -0.034$, p

Table 5.7 Robustness check: Alternative specifications to test role of OSS activity of acquirer

DV: Overall contributions (ln) Variables	(1) Acquirer's relative OSS activity (cat.)	(2) Acquirer's absolute OSS activity (cat.)	(3) Acquirer's LnCommits	(4) Dummy—integration subset
Time	−0.039*** (0.005)	−0.039*** (0.005)	−0.039*** (0.005)	−0.036*** (0.005)
PostAcquisition	−0.113 (0.069)	−0.114* (0.068)	−0.114* (0.069)	−0.180** (0.088)
Treated × PostAcquisition	−0.321*** (0.110)	−0.112 (0.131)	−0.110 (0.133)	–
Treated × PostAcquisition × AcquirerOSSActivity (categorized commits 1y prior acquisition *relative* to acquirer's size)	−0.040 (0.024)	–	–	–
Treated × PostAcquisition × AcquirerOSSActivity (categorized commits 1y prior acquisition—*absolute contributions*)	–	−0.034* (0.019)	–	–
Treated × PostAcquisition × AcquirerLnCommits	–	–	−0.029* (0.017)	–

(continued)

Table 5.7 (continued)

DV: Overall contributions (ln) Variables	(1) Acquirer's relative OSS activity (cat.)	(2) Acquirer's absolute OSS activity (cat.)	(3) Acquirer's LnCommits	(4) Dummy—integration subset
Treated × PostAcquisition × Acquirer active	-	-	-	−0.545*** (0.143)
Treated × PostAcquisition × Acquirer not active	-	-	-	−0.193 (0.163)
Constant	2.587*** (0.027)	2.601*** (0.026)	2.601*** (0.026)	2.633*** (0.029)
Project fixed effects	yes	yes	yes	yes
Observations	52,728	55,512	55,512	37,704
R²	0.565	0.570	0.570	0.567
Adj. R²	0.546	0.551	0.551	0.548
F-statistic	53.28***	55.99***	55.51***	49.26***

Note: The variable "treated" is not used as a covariate since it is captured in the project-level fixed effects. Cluster robust standard errors in parentheses. * p<0.10, ** p<0.05, *** p<0.01.

$= 0.075$; and Model 3: $\beta = -0.029$, $p = 0.082$). Lastly, we checked if the effect of more OSS-active acquirers leading to a stronger negative impact on OSS projects persists when we restrict the sample to acquisitions where integration took place. This is the case: The DiD estimator is -0.545 ($p = 0.001$) for OSS-active acquirers and -0.193 ($p = 0.177$) for non-OSS-active acquirers (Wald test: $p = 0.071$; Table 5.7, Model 4). Overall, our checks indicate robustness of our results also regarding the operationalization of the acquirer's OSS activity.

5.5 Qualitative Evidence

Our quantitative analysis reveals that on average OSS projects receive fewer contributions after an acquisition, and that this effect is more pronounced for projects where integration takes place, the acquirer itself is active in OSS, or projects that have a permissive license. All of these effects can be observed both for the target's and the community's contributions. Yet, the mechanisms that underlie these findings remain to be explored. For example, why are acquirers that themselves are active in OSS more harmful to the acquired target's OSS projects than acquirers without substantial OSS activities?

In order to uncover the mechanism driving our quantitative results, we conducted 52 semi-structured interviews with key decision-makers and experts in the industry. Most interviews were conducted online. In line with our exploratory research design, we conducted the majority of interviews (37/52) after the quantitative analysis to help us uncover the mechanisms behind our quantitative findings. The remaining interviews were conducted in an early stage of the research project to better understand the role of OSS in the context of acquisitions. We recorded and transcribed 48 interviews and took notes in four cases. We then coded the interviews using MaxQDA. Overall, our interviews lasted over 35h in total. Interviewing executives and developers working for the targets or acquirers as well as community members allowed us to capture the perspectives of all types of players involved in the acquisitions. Most of the interviews focused on projects that are also part of the sample used for the quantitative analysis. Interviewees were identified either by their GitHub or their LinkedIn profile. For seven acquisitions we were able to interview employees of the target as well as community members, giving us the chance to examine differences in the perception of the same acquisition across different types of stakeholders. The people we interviewed were distributed around the globe, covering North and South America, Europe, Asia, and Australia. Digital Appendix H provides an overview of exemplary codes along our clusters of codes.

5.5.1   Structural Integration

Our interviewees confirmed our finding that *structural integration* tends to have a negative impact on a target's contributions to OSS projects, and pointed to the underlying mechanisms. A senior engineer from a target informed us how it was a lot more difficult to contribute to OSS after being integrated in a large corporation: *"We were trying to meet [acquirer's] bureaucratic Open Source requirements by checking the dependencies, making sure they have licenses, getting into all that stuff. And it has been two years since we started the process and it hasn't happened yet."* Interviewees also describe that integration typically comes with a shift of developer resources to other tasks after the acquisition, to the neglect of the acquired target's OSS project. For example, a chief software architect of an OSS target integrated in a large tech company mentioned: *"We had a bunch of folks who were sort of directed into building a proprietary service on top of the Open Source technology. So those folks pivoted on the other side. [...] As well, given the nature of [acquirer], there were also a lot of opportunities. So, we saw some folks who were interested in going into a dev advocacy [bridge engineering team and community] or sort of moving around that way. So, there was just a natural kind of people migration."*

In case the target is *not* integrated into the acquirer's organization, the target's developer resources are typically allowed to continue its OSS development. When asked why the acquirer chose not to integrate, an acquired OSS developer mentioned: *"[Acquirer] seemed to be satisfied with what we were doing on our own. And well, sort of let us continue with the technology decisions we've made since there didn't seem to be a reason to rock the boat."* Later adding: *"The growth trajectory is what they focus on. So, it doesn't make sense to them to mess that up."* It became evident that acquirers who did not integrate the target typically acted as a friendly bystander supporting the target with financial resources, which can also be beneficial for the OSS work. An acquired software engineer explained: *"[...] for [target] it was great, because of course there was also a certain budget jump with [acquirer] to hire new employees [...] that way we can guarantee the further development of the product with new people."*

We note that the decision to integrate a target is endogenous, and likely contingent on the following two moderators, the acquirer's OSS activity and the license of the project.

5.5.2 Acquirer's OSS Activity

Our interviews also helped us understand why acquisitions by OSS-active acquirers entail a stronger reduction of contributions to the OSS project. OSS-active and not-OSS-active acquirers differ regarding if and how they exert control over OSS projects and the target's resources involved in them.

OSS-active acquirers have the ability to manage OSS developers and tend to shift the acquired employees to their own projects or let them integrate the acquired technologies with their existing products, hence draining resources from the focal public OSS project. An acquired software engineer reported: *"So, in that sense, some chunk of the team went off to work on [acquirer's new OSS project]. And a smaller part of the team was continuing, more or less, what they were doing before."* He also described the impact on the target's OSS project: *"So, people [...] saw the acquisition by [acquirer] as a threat, in the sense that the core [target's OSS project] work, for example would slow down or stop. Which, to some extent, it did. I mean obviously, some of the core team were now busy, working on [the acquirer's OSS project]."* A founder of an OSS-active acquirer underscored that acquirers may extract resources: *"We've had some smaller acquisitions in the past that were like that. It was more like acqui-hiring."* We also learned about two further mechanisms specific to OSS-active acquirers which lead to a negative impact on the acquired target's OSS projects. First, OSS-active acquirers let acquired developers integrate the acquired OSS technology with their own (OSS) stack, essentially slowing down the development of the acquired project. For example, one OSS-active acquirer had the target re-build the target's OSS project within their own OSS technology stack. Second, OSS-active acquirers stop projects that are duplicative to their own OSS efforts and redirect the resources to other projects as explained by one senior engineer: *"[The target] ended up shutting down [OSS project], over the space of the next six months or so, because it was duplicative of some stuff that [acquirer] had."*[19]

In contrast, not-OSS-active acquirers often aim to gain access to the OSS ecosystem, leaving their acquisition targets more freedom to continue their OSS projects. The CEO of a corporation that had acquired several OSS firms explained: *"If you're not familiar with Open Source, the best play is the long play—expecting that the returns will come later than you want so that you can ensure the ecosystem is not disrupted. As soon as it gets disrupted, you'll find that people will abandon*

---

[19] In this case, the employees were moved to other OSS projects, building proprietary software on top of other OSS projects or completely changed their role within the acquirer. Most employees stayed with the acquirer.

the project. So, I think it's really, really critical that things work out that way."
Acquirers not active in OSS often following this approach results in OSS projects
of targets acquired by such firms being less negatively affected by acquisitions.
A developer working for an acquired sponsoring firm described the motivation of
such acquirers: *"They [the acquirer] wanted to kind of learn how to become a part
of this community."*

Taken together, our interviews reveal the mechanism behind the more pronounced decline of OSS projects subsequent to acquisitions when acquirers themselves
are active in OSS: Only those acquirers have the ability to integrate OSS developers into their own software development, and also have alternative uses for
those resources in their own projects. As a result, they extract more resources
from the target's OSS projects.

5.5.3   Role of Licenses

The qualitative evidence also helps to understand why permissive licensed projects are more negatively affected by acquisitions than copyleft licensed projects:
acquirers can exert more control over those projects, and resource extraction is
easier. As one community member explained: *"When an MIT licensed project [the
interviewee later specified 'permissive licenses, like MIT or BSD'] is acquired, the
possibility of a closed source fork arises. [...] Closed source fork means that the new
owner continues the project closed source."* In contrast, copyleft licensed projects,
which require derivative work to be under the same license as the original work,
exclude use of the software in proprietary software development thereby restric-
ting the acquirer's possibilities to exert control (unless all copyright holders agree,
which is typically difficult to achieve for community-based projects). The founder
of an acquisition target, who was dissatisfied with how the acquirer handled the
target's OSS projects after the acquisition, underscored this point: *"[Before the
acquisition] I was always in favor of MIT and permissive and now I am more or
less [in favor of the copyleft-type] LGPL because I say, it is important to me that
it remains free and that [acquirers] do not take it and take advantage of it. [...]
They [acquirer] can do what they want [with permissive licensed projects]."* We
also learned that resource extraction from copyleft licensed projects is more difficult for acquirers: Employees and the community of copyleft licensed projects
are more likely to resist interference by the acquirer, and as a result the acquirer prefers a hands-off approach. The CEO of an OSS-active acquirer explained,

"with the GPL, you typically have a very, very rabid religious Open Source community. So, you've got to be really, really careful with those guys. Because they expect everything that their project, that they've been part of, touches to be GPL."

5.5.4 Mechanisms Linking Acquisition and Community Response

Our interviews also clarified how communities react to acquisitions, explaining our results in Model 5 in our main quantitative analysis, where we found that changes in contributions from the community after acquisitions can largely be explained by lagged changes of the contribution behavior of the target. Our interviewees generally stated that communities do not react particularly strongly to acquisitions per se, in some cases not even being aware of them. Rather, community members respond to the behavior of key employees. As a community member explained: *"When I see something like this being acquired, I'm always concerned. But this has changed over the years. Initially, you tended to see kind of an embrace, extend, and extinguishment of Open Source projects that were acquired. So, my first reaction used to be very, very pessimistic. [...] But I think that companies have become much more aware of why they would participate in Open Source and what people really need in order for an Open Source project to remain viable. So, my approach has become more of a wait and see approach. So, if I see an acquisition then I keep a lookout to: Are core developers hired? Do the repositories remain accessible? Is the roadmap impacted in a major way? So, I look more at how it's being run in practice."* Acquirers' are well aware of this "wait and see" approach by the community. The CTO of an acquirer highlighted how they approached the community: *"I think most of the time people are scared of what you're going to do and if you're going to shut it down. And so that is the big fear. And so really a lot of it comes down to making sure that they are safe."* When asked how they made the community feel safe, he replied: *"Just talking about it all the time and then just showing them by example. A lot of just repeating over and over and over again: We are investing in the technology; we're not shutting it down."* Interestingly, from email exchanges we learned that community members often did not even notice acquisitions where no integration took place—a point that highlights the critical role of the extent to which an acquirer exerts control over targets also for the reaction of the community.

5.5.5   Alternative Explanation: Acquisitions to Eliminate OSS Competition

We have suggested that acquisitions lead to a reduction in contributions when the acquirer extracts resources from the target and the OSS project and that it does so because it can deploy these resources more effectively elsewhere. An alternative explanation could be that the acquirer seeks to "kill" the OSS project because it competes with its own offerings—similar to the phenomenon of "killer acquisitions" as identified by Cunningham et al. (2020) where incumbents acquire targets to discontinue the target's innovation projects and pre-empt future competition by potential substitutes. Our qualitative evidence does, however, suggest that in cases where the target's OSS project was discontinued the acquisition of resources was a primary motive. Only in one case an interviewee suspected a competitive threat by an OSS project as an acquisition motive, among others. A target employee stated: *"[…] it was basically an acqui-hire of the team. Mainly. So, there was that, and some combination of [acqui-hire with] them seeing [target] as a threat in some way."* The CEO of the (OSS-active) acquirer, however, denied the motive of shutting down a competing OSS project: *"We were always acquiring for engineering talent and or product augmentation."* Of course, this comes with the caveat that the involved actors may be unwilling to reveal to us (and themselves) that an acquisition did indeed constitute a killer acquisition.

5.6   Discussion and Conclusion

We studied the impact of acquisitions on contributions to OSS projects. While we cannot claim to fully prove causality, our DiD approach combined with coarsened exact matching and interviews suggests that what we observe are indeed the effects of acquisitions. Our use of the terms "effect" and "impact" should be understood with this caveat in mind when we summarize our findings in the following.

5.6.1   Summary of Findings

On average, acquisitions have a negative impact on contributions to OSS projects. Disentangling the overall effect into contributions from the target firm and those from the community, we find a stronger impact on the former. The community seems to be primarily reacting to how the target's activities within the respective project change.

An analysis of moderating variables reveals that the negative impact of an acquisition on contributions from both target and community is stronger if the target is structurally integrated into the acquirer's organization, if the acquirer is itself active in OSS development, and if the respective project is governed by a permissive license. We note that all three contingencies relate to the acquirer's tendency and ability to extract resources from the target and its OSS projects: Higher tendency and ability of the acquirer to extract resources from the target tend to be harmful to the project. We suggest that our study relates to four literature streams.

5.6.2 Contribution

Survival and success of OSS projects. Our primary contribution is to OSS research, where we respond to Shah and Nagle's (2020) call for more empirical research on the role of a firm's engagement for the survival and success of communities. Our study helps to understand the factors that shape contributions to OSS, and thus determine the survival and success of OSS projects (Fang and Neufeld, 2009; Ho and Rai, 2017; Shah, 2006). While our finding on the on average damaging effect of the acquisition of a sponsoring firm informs research on the survival and success of OSS projects by itself, the variation among acquisitions leads to more broadly applicable insights. We find to our own surprise that acquirers that are themselves active and experienced in OSS development tend to have a more harmful effect on OSS projects, showing that a change in control over a project to an acquirer itself experienced and familiar with OSS is no guarantee for the survival and success of a target's project. In fact, our interviews show that particularly such firms tend to have valuable alternative opportunities to use the target's resources engaged in OSS—rendering the extraction of resources more likely—and thus endanger the success and survival of the OSS project. On a higher level, these findings highlight the importance of considering the context of an OSS project in order to understand, predict, and manage its development.

Our findings regarding the role of the deployed license for a project's robustness informs research on the governance of OSS projects (He et al., 2020; Lerner and Tirole, 2005), illustrating that licenses can help to protect the survival and success of the OSS project when the controlling actors change. We show that a higher degree of freedom, such as provided by permissive licenses compared to copyleft licenses, renders a project more vulnerable to acquirers aiming to extract resources from the target. This illustrates that formal governance mechanisms such as licenses become particularly important when the controlling actors in an OSS project change.

Firm activities and benefits in OSS. Our findings on the ability and tendency to extract resources also inform our understanding of firms' OSS strategies and the benefits they derive from engaging in OSS. Our study reveals that firms acquiring sponsoring firms pursue *two alternative types of strategies* that benefit them in alternative ways: they either extract resources from the acquired sponsoring firm, effectively harming the OSS project, or they let the target continue to contribute to the OSS project, with the effect of preserving it. We also illustrate under what kind of conditions firms tend to choose one or the other strategy. We illustrate that when firms have a background in OSS themselves and/or the license of the OSS projects allows them to do so, they tend to choose the first "resource extraction" strategy. We also show that a potential alternative explanation of the finding that acquisitions on average lead to the reduction of contributions—i.e., the motive to kill competitive OSS projects—only plays a minor role for the case of targets active in OSS development. Our interviews showed that in cases where the target's OSS project was discontinued, acquirers mostly seek the targets' resources.

Management of external resources. Third, we contribute to studies on the management of external resources such as OSS and open innovation communities. While existing research has focused on how to govern and manage these resources (Dahlander and Magnusson, 2005; Dahlander and Wallin, 2006; Franke et al., 2013; Klapper and Reitzig, 2018; West and O'Mahony, 2008), we provide first insights about the *transferability* of external resources. Our study suggests that communities—individuals typically working on an OSS project for other reasons than the project sponsor—do not mainly react to a transfer of control over a project per se, but rather to the post-acquisition behavior of the acquirer and the target. Instead of the acquisition itself having a strong effect on the contributions by the community, the community's contributions appear to be contingent on the sponsor's engagement and on the established direction and purpose of the project. Thus, an acquirer that seeks to maintain the target's external resources should refrain from extracting resources from the project, from shifting the direction of the project, and from stripping the community off its power to influence the direction of the OSS project.

Level of integration. Finally, our study adds new aspects to the debate on the effects of different levels of integration in the literature on technology acquisitions. Studying the advantages and drawbacks of a tighter or earlier integration of the acquisition target, scholars have identified a number of contingencies such as the acquisition being related vs. unrelated (Datta and Grant, 1990), the level of engagement of acquired managers (Graebner, 2004), the developmental stage of the target's technology (Puranam et al., 2006), the previous acquisition experience of the acquirer (Choi and McNamara, 2018), and complementarity vs. similarity of resources (Zaheer, Castañer, and Souder, 2013). Our research adds the aspect of external resources, in our case, an OSS community: Tighter integration of the target by the acquirer, the extraction of resources from the project, and changes in direction of the collaboration appear to reduce the availability of the external resources.

5.6.3   Managerial Implications, Limitations, and Outlook

Managerial implications. Our findings have a number of managerial implications. Founders who consider selling their OSS firm and wish to keep the related OSS project alive should be aware that acquirers that envisage organizational integration of the target as well as, counterintuitively, OSS-active acquirers are likely harmful to the project. Founders may also consider, when initiating an OSS project, to choose a copyleft-type license—although, of course, this choice may diminish the odds of an acquisition in the first place. Similarly, potential acquirers should be aware of the adverse effects that an acquisition may have on OSS projects. In particular, they should know that the potential to extract resources from the project or to steer it into a new direction are limited if the project shall continue to prosper. Finally, members of an OSS community—hobbyists as well as employees of other firms—can use our results to assess the likely impact of an acquisition on the project, and can make their participation choice contingent on it.

Limitations. Our study is subject to several limitations. Despite our efforts to come closer to causal inference with matching and DiD analysis, there may still be a selection bias regarding which targets get acquired and are willing to be acquired. While it appears somewhat implausible that acquirers seek out targets whose OSS projects they expect to decline after the acquisition date, it might be that a potential target is more willing to be acquired if they were planning to withdraw resources from their OSS project anyway. Yet, our interviews provided no evidence in this regard.

Also, we lack quantitative evidence to what extent the drop in contributions to a target's OSS projects is caused by a shift of resources to other projects or a decline of resource productivity due to the integration of the target. While our interviews provide strong evidence that resource extraction is a main driver of the drop in contributions to a target's projects and several interviewees explicitly declined that they experienced a loss in productivity, we cannot fully rule out the latter option. A more nuanced quantitative understanding about how resources are utilized after an acquisition could help to further understand the mechanisms at play.

Lastly, the use of commits as a measure of contribution activity in OSS projects has—while regularly being used in management and information systems research (e.g., Daniel et al., 2018; Nagle, 2019)—some limitations. Particularly, the "value" and complexity of commits may vary strongly. Other measures, like the lines of code or the complexity of code, both difficult to obtain for a large number of projects as in our study, could be tested in additional studies.

Outlook. We believe that this study helps elucidate central questions about the role of acquisitions for the evolution and management of OSS projects and communities. It is important to further extend this line of inquiry in order to better understand the role of acquisitions in the context of open innovation. In this study we have examined the consequences of acquisitions in the context of OSS. Future research may study the antecedents. For example, how does a firm's engagement in OSS affect its chances of being acquired and by whom it will be acquired. Similarly, being active in open innovation might influence the acquirer's target sourcing and evaluation processes. Furthermore, recent work on acquisitions has shown how networks are shaped by acquisitions (Hernandez and Shaver, 2019). The work on acquisitions in the context of OSS differs in an interesting manner as the project, not just the firm, is embedded in a web of social and technical interdependencies. Future research could examine whether acquirers seek a particular network position and how stakeholders in other projects respond if a project they depend on is acquired.

Summary and Outlook

6

The goal of this thesis was to explore the intersection of OSS development and acquisitions. Despite the economic and practical importance of OSS and the increasing number of acquisitions in the OSS, research has so far overlooked the intersection of those research streams. To do so, this thesis took two perspectives on the role of OSS in acquisitions. The first perspective, covered in Chapters 3 and 4, aimed to understand the role of OSS development as an input to the strategic decision-making process of an acquirer for acquisitions. The second perspective aimed to understand the impact of acquisitions on OSS projects, contributions to those from firms and the communities around them, and contingencies thereof.

6.1   Findings and Contribution

The findings of this thesis suggest that both perspectives on OSS development and acquisitions are relevant. OSS development can play a role in the decision-making process leading to acquisitions, and OSS development can be influenced by acquisitions. In the following, I will briefly summarize the key findings and contributions from each chapter.

Chapter 3 aimed to qualitatively generate first evidence of whether and how a target's OSS activities are relevant for it being acquired. To do so, I conducted 52 interviews with acquirers, targets, and community members, allowing me to holistically examine the role of OSS development in acquisitions. I find that some, not all, targets do indeed get acquired because of their OSS activities. More specifically, I found six different OSS-related acquisition motives related to a target's OSS talent, its OSS community, its brand in OSS, the fast adoption of its OSS technology or product, its OSS culture, and the potential of its OSS to become a competitive threat for the acquirer. Furthermore, I found that a target's activities

M. Vetter, *Acquisitions and Open Source Software Development*, Innovation und Entrepreneurship, https://doi.org/10.1007/978-3-658-35084-0_6

in OSS can be relevant for an acquirer's target search, evaluation, and selection processes. Being active in OSS can make a firm visible to acquirers, and a firm's activities in OSS can be a valuable input for the target evaluation and selection. Lastly, interviews suggested that OSS licenses employed by a firm are a particularly relevant aspect when it comes to it being acquired. These findings contribute to different streams of literature. First, they extend research on acquisition motives (Chatterjee, 1986; Cunningham et al., 2020; Grimpe and Hussinger, 2008; Hitt, Hoskisson, Johnson, and Moesel, 1996; Ouimet and Zarutskie, 2012; Seth, 1990; Walter and Barney, 1990; Worek et al., 2018) by adding additional motives related to OSS activities to the list of known acquisition motives. Second, they add to the research on information asymmetry between target and acquirer (Capron and Shen, 2007; Hussinger, 2010; Shen and Reuer, 2005) by showing that OSS activity of the target can reduce information asymmetry between target and acquirer as the quality of OSS code and talent and the interaction of the target with the community are better observable than in proprietary software development. Lastly, the findings add new aspects to research on acquisitions in markets for technology (Arora et al., 2001). Specifically, they add the influence of the target's choice of licenses for its IP on the acquirer's decision to acquire the target and the decision-making process leading to this decision.

Chapter 4 added to the perspective on the role of OSS development in the pre-acquisition phase. Specifically, I quantitatively examined if and how targets' and acquirers' OSS activities before an acquisition influence acquisition likelihood and acquisition timing. The main data sources I used for my quantitative analyses are Crunchbase (acquisition data) and GitHub (OSS development data). Using survival and competing risk models, I found that firms active in OSS development are associated with a higher acquisition likelihood than firms not active in OSS development. Additionally, within the group of targets active in OSS development, acquirers themselves active in OSS development are associated with significantly earlier acquisitions than acquirers not active in OSS. While I could not fully claim causality in my analysis, I discussed the potential mechanisms behind those findings in light of prior research and my findings from the interviews. Potential reasons for the higher acquisition likelihood of firms active in OSS are the better visibility of firms active in OSS, the potential to better evaluate them, and the potential that being active in OSS allows firms to create better products and do so faster than when developing proprietary software. I ruled out other potential mechanisms, such as a hype of OSS or the need of owners of firms active in OSS to more often sell their firms due to problems in monetizing the OSS. Regarding the role of an acquirers OSS activity for the timing of acquisitions, I identified two potential mechanisms. The effect of acquirers themselves active in

OSS development being associated with earlier acquisitions of targets active in OSS development might be driven by their better access to such targets and/or their better capabilities to evaluate them. These findings contribute to different streams of literature. First, it expands the understanding of firm characteristics, influencing their attractiveness as a target and their likelihood of getting acquired (Chakrabarti and Mitchell, 2013; Fischer et al., 2020; Hernandez and Shaver, 2019; Ransbotham and Mitra, 2010; Rogan and Sorenson, 2014) by showing that firms active in OSS are associated with a higher acquisition likelihood. Second, the findings contribute to research aiming to understand the role of an acquirer's capabilities for the timing of acquisitions (Hlavka, 2019) by suggesting that acquirers can build distinct capabilities in finding and evaluating targets active in OSS leading to earlier acquisitions. Lastly, the findings contribute to research on the strategic dimensions of a firm's openness decision (Chesbrough, 2006; Dahlander and Gann, 2010) by showing how openness can influence strategic decision making in acquisition decision-making processes.

Chapter 5 focused on the impact of acquisitions on OSS development. Using a mixed-methods approach combining quantitative analysis based on Crunchbase and GitHub data and results from the aforementioned 52 interviews, I, together with my co-authors Joachim Henkel (TUM) and Henning Piezunka (INSEAD), covered the impact of acquisitions on firms' activities in OSS projects as well as the impact on the community around them. Furthermore, contingencies across acquisitions and the drivers behind those contingencies were covered in this chapter. We found that, on average, acquisitions have a negative impact on contributions to OSS projects. Disentangling the overall effect into contributions from the target firm and those from the community, we found a stronger impact on the former. The community seems to be primarily reacting to how the target's activities within the respective project change. An analysis of moderating variables revealed that the negative impact of an acquisition on contributions from both target and community is stronger if the target is structurally integrated into the acquirer's organization, if the acquirer is itself active in OSS development, and if the respective project is governed by a permissive license. Combining these quantitative findings with findings from the interviews, we suggest that all three contingencies relate to the acquirer's tendency and ability to extract resources from the target and its OSS projects: Higher tendency and ability of the acquirer to extract resources from the target tend to be harmful to the project. These findings contribute to different streams of literature. First, they contribute to research on the factors determining the survival and success of OSS projects (Fang and Neufeld, 2009; Ho and Rai, 2017; Shah, 2006) by showing that acquisitions, on average, have a negative impact on the survival and success of OSS projects, that acquirers themselves active and experienced in OSS development tend to have a more harmful effect on OSS projects due to their increased willingness and ability to extract resources from the target's OSS projects, and that license choice can help to protect the survival and success of the OSS project when the controlling actors change. Second, the findings contribute to research on the management of external resources, such as communities (Dahlander and Magnusson, 2005; Dahlander and Wallin, 2006; Franke et al., 2013; Klapper and Reitzig, 2018; West and O'Mahony, 2008) by providing first insights about the transferability of such external resources. The study in this chapter suggests that communities do not mainly react to a transfer of control over a project per se, but rather to the post-acquisition behavior of the acquirer and the target. Instead of the acquisition itself having a strong effect on the contributions by the community, the community's contributions appear to be contingent on the acquirer's engagement and on the established direction and purpose of the project after an acquisition. Lastly, the findings contribute to research on the level of structural integration of targets (Choi and McNamara, 2018; Datta and Grant, 1990; Graebner, 2004; Puranam et al., 2006; Zaheer et al., 2013) by shedding light on the impact of structural integration on external resources such as communities. Tighter integration of the target by the acquirer, the extraction of resources from the project, and changes in the direction of the collaboration appear to reduce the availability of the external resources.

6.2   Practical Implications

The presented research in this thesis also has several implications for practitioners. The findings regarding the role of OSS activities for the decision-making process leading to acquisitions are relevant for founders and their investors. If they consider selling their firm as a promising exit strategy, they need to be aware of the potential influence of their firm's OSS activities on the acquirer's perception of the firm as a potential target and how acquirers will evaluate their OSS activities. Participating in OSS development seems not only useful for attracting talent but also for showcasing (OSS) capabilities to potential acquirers and thus increasing the probability of getting acquired.

The findings regarding the role of OSS activities for the decision-making process leading to acquisitions are also relevant for acquirers. Acquirers need to be aware of the potential OSS development of potential targets offers for reducing technology, market, and integration risks associated with acquisitions. However, they need to consider that in order to be able to exploit this potential, acquirers need to be knowledgeable about OSS development themselves, which typically requires acquirers to be active in OSS at least to a certain level prior to the acquisition. Furthermore, being active in OSS themselves may enable them to become aware of other market participants and potential targets faster and allow them to better evaluate a potential target's quality leading to better acquisition decisions and earlier acquisitions.

The findings regarding the impact of acquisitions on OSS development are relevant for acquirers, managers and owners of targets, and community members. Owners and managers who consider selling their firm and wish to keep the related OSS project alive should be aware that acquirers that envisage organizational integration of the target are likely harmful to the target's OSS projects. When initiating an OSS project, they may also consider choosing a copyleft-type license if they want to protect their OSS project from the impact of a future acquisition. Acquirers should be aware of the adverse effects that an acquisition may have on OSS projects. While it is in some way good news for acquirers that communities are not inherently opposing acquisitions, they should know that the potential to extract resources from the project or to steer it into a new direction is limited if the project shall continue to prosper. Finally, members of an OSS community— hobbyists as well as employees of other firms—can use the results to assess the likely impact of an acquisition on the project and can make their participation choice contingent on it.

6.3   Limitations and Future Research

As detailed in Chapters 3, 4, and 5, this thesis is not free from limitations, which opens up some potential for future research. Without repeating all limitations here, I want to point out some key limitations across the chapters. Furthermore, I want to highlight additional avenues for future research.

First, the transferability of the findings to other types of open innovation and earlier years of the OSS movement is limited, as I selected the sample of acquisitions by searching for acquired firms active on GitHub. Focusing on GitHub only limits the scope of the observations underlying the findings in this thesis, as other platforms for OSS development, such as SourceForge exist. Furthermore, GitHub data is only available after 2010. While I do not foresee many differences between OSS development on GitHub compared to other platforms, the years covered should be taken into account when transferring the given findings to other acquisitions in earlier years. In the early years of OSS development, many firms were much more sceptical about OSS development, which might imply a different view of acquirers towards potential targets active in OSS development.

Second, there are some methodological limitations in the quantitative parts of this thesis. Despite my effort to come close to causal inference by using matching and constructing several robustness checks, my choice of methods and limitations from the available data do not allow me to fully prove causal relations quantitatively. In particular, in Chapter 4 I am not able to quantitatively differentiate the effects of different mechanisms, which would all result in the observed main effect of an increased acquisition likelihood of firms active in OSS development. In Chapter 5, there may still be a selection bias regarding which targets get acquired, and are willing to be acquired. While it appears somewhat implausible that acquirers seek out targets whose OSS projects they expect to decline after the acquisition date, it might be that a potential target is more willing to be acquired if they were planning to withdraw resources from their OSS project anyway. A more nuanced quantitative understanding of the motives in acquisitions and the mechanisms at play could help to further understand the phenomenon of acquisitions of targets active in OSS development.

Lastly, the use of commits to identify firms active in OSS development has—while regularly being used in management and information systems research (e.g., Daniel et al., 2018; Nagle, 2019)—some limitations. Firms might also participate in OSS by being active in mailing lists, forums, or at events and could technically—while unlikely—run an OSS project without actively contributing to the project themselves. Furthermore, the "value" and the complexity of commits may vary strongly. Other measures, like the lines of code or the complexity of code, could be tested in additional studies.

I believe that this thesis helps elucidate central questions about the phenomenon of acquisitions of firms active in OSS development. It is important to conduct further research on this increasingly important phenomenon to better understand the role of acquisitions for open innovation and the role of open innovation for acquisitions. For example, recent work on acquisitions has shown how networks are shaped by acquisitions (Hernandez and Shaver, 2019). The work on acquisitions in the context of OSS differs in an interesting manner as the OSS project, not just the firm, is embedded in a web of social and technical interdependencies. Future research could examine whether acquirers seek a particular network position, how a firm's network position influences its acquisition likelihood, and how stakeholders in other projects respond if a project they depend on is acquired. Another interesting avenue for research could be to further differentiate the type of involvement of firms in OSS development. While this thesis exclusively focused on firms actively contributing to OSS, it would be interesting to see if the findings from this thesis are also valid for firms only using OSS—or if there are significant differences. Lastly, this thesis almost exclusively focused on the impact of acquisitions on targets and their communities. An interesting avenue of research could be the impact of such acquisitions on acquirers. Particularly if acquirers with no prior experience in OSS development themselves can successfully integrate the targets knowledge and OSS culture into their own organization is relevant here.

Taken together, this thesis provides initial answers to key questions about the phenomenon of acquisitions of firms active in OSS development. However, many questions remain unaddressed. The phenomenon of acquisitions of firms active in OSS development provides a rich source and many avenues for further investigation. To conclude, I hope the research presented in this thesis demonstrates that the intersection of research on OSS development and acquisitions can be a rich source of empirical and theoretical knowledge and a playing field for future research activity.

Bibliography

Agarwal, R., Braguinsky, S., & Ohyama, A. (2019). Centers of gravity: The effect of stable shared leadership in top management teams on firm growth and industry evolution. *Strategic Management Journal, 41*(3), 1–32.

Ågerfalk, P. J., & Fitzgerald, B. (2008). Outsourcing to an unknown workforce: Exploring opensourcing as a global sourcing strategy. *Management Information Systems Quarterly, 32*(2), 385–409.

Aghasi, K., Colombo, M. G., & Rossi-Lamastra, C. (2017). Acquisitions of small high-tech firms as a mechanism for external knowledge sourcing: The integration-autonomy dilemma. *Technological Forecasting & Social Change, 120*, 334–346

Ahuja, G., & Katila, R. (2001). Technological Acquisitions and the Innovation Performance of Acquiring Firms: A Longitudinal Study. *Strategic Management Journal, 22*(3), 197–220.

Aktas, N., de Bodt, E., & Cousin, J.-G. (2007). Event studies with a contaminated estimation period. *Journal of Corporate Finance, 13*(1), 129–145.

Alexy, O. (2009). *Free Revealing: How Firms Can Profit From Being Open.* Wiesbaden, Germany: Gabler Verlag.

Alexy, O., Block, J., Sandner, P., & Ter Wal, A. (2012). Social capital of venture capitalists and start-up funding. *Small Business Economics 39*, 835–851.

Alexy, O., George, G., & Salter, A. J. (2013). Cui Bono? The selective revealing of knowledge and its implications for innovative activity. *Academy of Management Review, 38*(2), 270–291.

Alexy, O., West, J., Klapper, H., & Reitzig, M. (2017). Surrendering control to gain advantage: Reconciling openness and the resource-based view of the firm. *Strategic Management Journal, 39*(6), 1704–1727.

Alexy, O., Frederiksen, L., & Hutter, K. (2020). Quo Vadis, open and user innovation theory?, *Innovation, 22*(2), 97–104.

Allain, M.-L., Henry, E., & Kyle, M. (2016). Competition and the Efficiency of Markets for Technology. *Management Science, 62*(4), 1000–1019.

Almirall, E., & Casadesus-Masanell, R. (2010). Open versus closed innovation: A model of discovery and divergence. *Academy of Management Review, 35*(1), 27–47.

Alvarez, L. H. R., & Stenbacka, R. (2006). Takeover timing, implementation uncertainty, and embedded divestment options. *Review of Finance, 10*(3), 417–441.

Anand, J., & Singh, H. (1997). Asset redeployment, acquisitions and corporate strategy in declining industries. *Strategic Management Journal, 18*(S1), 99–118.

Andersen-Gott, M., Ghinea, G., & Bygstad, B. (2012). Why do commercial companies contribute to open source software? *International Journal of Information Management, 32*(2), 106–117.

Angwin, D. (2004). Speed in M&A Integration. *European Management Journal, 22*(4), 418–430.

Angwin, D. (2007). Motive Archetypes in Mergers and Acquisitions (M&A): The Implications of a Configurational Approach to Performance. In: C.L. Cooper, & S. Finkelstein (Eds.), *Advances in Mergers and Acquisitions Vol. 6* (pp. 77–105), Bingley, UK: Emerald Group Publishing Limited.

Angwin, D. N., Paroutis, S., & Connel, R. (2015). Why good things Don't happen: the microfoundations of routines in the M&A process. *Journal of Business Research, 69*(6), 1367–1381.

Arora, A., Fosfuri, A., & Gambardella, A. (2001). Markets for Technology and their Implications for Corporate Strategy. *Industrial and Corporate Change, 10*(2), 419–451.

Arora, A., Belenzon, S., & Sheer, L. (2017). Back to Basics: Why do Firms Invest in Research? NBER working paper series, No. 23187. Cambridge, MA: National Bureau of Economic Research.

Arvanitis, S., & Stucki, T. (2014). How Swiss small and medium-sized firms assess the performance impact of mergers and acquisitions. *Small Business Economics, 42*(2), 339–360.

Bagchi, P., & Rao, R. P. (1992). Decision making in mergers: an application of the analytic hierarchy process. *Managerial and Decision Economics, 13*(2), 91–99.

Baldwin, C., & von Hippel, E. (2011). Modeling a Paradigm Shift: From Producer Innovation to User and Open Collaborative Innovation. *Organization Science, 22*(6), 1399–1417.

Balka, K., Raasch, C., & Herstatt, C. (2014). The effect of selective openness on value creation in user innovation communities. *Journal of Product Innovation Management, 31*(2), 392–407.

Bamberger, P. A. (2018). Clarifying What We Are about and Where We Are Going. *Academy of Management Discoveries, 4*(1), 1–10.

Bannert, V., & Tschirky, H. (2004). Integration Planning for Technology Intensive Acquisitions. *R&D Management, 34*(5), 481–494.

Bansal, P., & Corley, K. (2011). From the Editors—The Coming of Age for Qualitative Research: Embracing the Diversity of Qualitative Methods. *Academy of Management Journal, 54*(2), 233–237.

Barkema, H. G., & Schijven, M. (2008). How do firms learn to make acquisitions? A review of past research and an agenda for the future. *Journal of Management, 34*(3), 594–634.

Barney, J. B. (1988). Returns to bidding firms in mergers and acquisitions: reconsidering the relatedness hypothesis. *Strategic Management Journal, 9*(S1), 71–78.

Bateman, P. J., Gray, P. H., & Butler, B. S. (2011). The impact of community commitment on participation in online communities. *Information Systems Research, 22*(4), 841–854.

Bauer, F., & Matzler, K. (2014). Antecedents of M&A Success: The Role of Strategic Complementarity, Cultural Fit, and Degree and Speed of Integration. *Strategic Management Journal, 35*(2), 269–291.

Bauer, F., Matzler, K., & Wolf, S. (2016). M&A and innovation: The role of integration and cultural differences— A central European targets perspective. *International Business Review, 25*(1), 76–86.

Behlendorf, B. (1999). Open Source as a Business Strategy. In C. DiBona, S. Ockman, & M. Stone (Eds.), *Open Sources: Voices from the Open Source Revolution* (pp. 149–170). Sebastopol, CA: O'Reilly & Associates.

Benson, C. Müller-Prove, M., & Mzourek, J. (2004). Professional usability in open source projects: GNOME, OpenOffice.org, NetBeans. *Extended abstracts on Human factors in computing systems* (pp. 1083–1084), CHI 2004, Vienna, Austria.

Bertrand, O. (2009). Effects of foreign acquisitions on R&D activity: Evidence from firm-level data for France. *Research Policy, 38*(6), 1021–1031.

Bergquist, M., & Ljungberg, J. (2001). The Power of Gifts: Organising Social Relationships in Open Source Communities. *Information Systems Journal, 11*(4), 305–320.

Birkinshaw, J., Bresman, H., & Hakanson, L. (2000). Managing the post acquisition integration process: How the human integration and task integration. *Journal of Management Studies, 37*(3), 395–425.

Bettinazzi, E.L., & Zollo, M. (2017). Stakeholder Orientation and Acquisition Performance. *Strategic Management Journal, 38*(12), 2465–2485.

Black Duck Software (2016). *The Eighth Annual Future of Open Source Survey.* Retrieved from https://de2.slideshare.net/blackducksoftware/2016-future-of-open-source-survey-results on 16.06.2020.

Blackwell, M., Iacus, S., King, G., & Porro, G. (2009). CEM: Coarsened exact matching in Stata. *Stata Journal, 9*(4), 524–546.

Block, J. H., Fisch, C. O., Hahn, A., & Sandner, P. G. (2015). Why do SMEs file trademarks? Insights from firms in innovative industries. *Research Policy, 44*(10), 1915–1930.

Bonaccorsi, A., Giannangeli, S., & Rossi, C. (2006). Entry strategies under competing standards: Hybrid business models in the open source software industry. *Management Science, 52*(7), 1085–1098.

Borges, H., & Valente, M. T. (2018). What's in a GitHub Star? Understanding Repository Starring Practices in a Social Coding Platform. *Journal of Systems and Software, 146,* 112–129.

Bower, J. L. (2001). Not all M&A are alike—And that matters. *Harvard Business Review, 79,* 93–101.

Bradburn, M. J., Clark, T. G., & Love, S. B. (2003). Survival Analysis Part II: Multivariate data analysis—an introduction to concepts and methods. *British Journal of Cancer, 89*(3), 431–436.

Brandenburger, A. M., & Stuart, H. W. (1996). Value-based business strategy. *Journal of Economics and Management Strategy, 5*(1), 5–24.

Bryant, A., & Charmaz, K. (2007). *The SAGE Handbook of Grounded Theory*, London, UK: SAGE Publications.

Burton, R. M., Håkonsson, D. D., Nickerson, J., Puranam, P., Workiewicz, M., & Zenger, T. (2017). GitHub: Exploring the space between boss-less and hierarchical forms of organizing. *Journal of Organization Design, 6*(1), 1–19.

Capra, E., Francalanci, C., Merlo, F., & Rossi-Lamastra, C. (2011). Firms' Involvement in Open Source Projects: A Trade-Off between Software Structural Quality and Popularity, *Journal of Systems and Software, 84*(1), 144–161.

Capron, L., Dussauge, P., & Mitchell, W. (1998). Resource redeployment following horizontal acquisitions in Europe and North America, 1988–1992. *Strategic Management Journal, 19*(7), 631–661.

Capron, L., Mitchell, W., & Swaminathan, A. (2001). Asset divestiture following horizontal acquisitions: A dynamic view. *Strategic Management Journal, 22*(9), 817–844.

Capron, L., & Mitchell, W. (2009). Selection capability: How capability gaps and internal social frictions affect internal and external strategic renewal. *Organization Science, 20*(2), 294–312.

Capron, L., & Shen, J. C. (2007). Acquisitions of private versus public firms: private information, target selection and acquirer returns. *Strategic Management Journal, 28*(9), 891–911.

Carow, K., Heron, R., & Saxton, T. (2004). Do early birds get the returns? An empirical investigation of early-mover advantages in acquisitions. *Strategic Management Journal, 25*(6), 563–585.

Cassiman, B., Colombo, M., Garrone, P., & Veugelers, R. (2005). The impact of M&A on the R&D process. An empirical analysis of the role of technological and market relatedness. *Research Policy 34*(2), 195–220.

Cassiman, B., & Veugelers, R. (2006). In Search of Complementarity in Innovation Strategy: Internal R&D and External Knowledge Acquisition. *Management Science, 52*(1), 68–82.

Chakrabarti, A., & Mitchell, W. (2013). The persistent effect of geographic distance in acquisition target selection. *Organization Science, 24*(6), 1805–1826.

Chakrabarti, A., Hauschildt, J., & Sueverkruep, C. (1994). Does it pay to acquire technological firms? *R&D Management, 24*(1), 47–56.

Chan, J., & Husted, K. (2010). Dual Allegiance and Knowledge Sharing in Open Source Software Firms. *Creativity and Innovation Management, 19*(3), 314–326.

Charmaz, K. (2006). *Constructing Grounded Theory —A Practical Guide through Qualitative Analysis*, Thousand Oaks, CA: SAGE Publications.

Chatterjee, S. (1986). Types of synergy and economic value: The impact of acquisitions on merging and rival firms. *Strategic Management Journal, 7*(2), 119–139.

Chatterji, A., & Patro, A. (2014). Dynamic Capabilities and Managing Human Capital. *Academy of Management Perspectives, 28*(4), 395–408.

Chaudhuri, S., & Tabrizi, B. (1999). Capturing the Real Value in High-Tech Acquisitions. *Harvard business review, 77*(5), 123–130.

Chaudhuri, S., Marco, I., & Tabrizi, B. N., (2005). The Multilevel Impact of Complexity and Uncertainty on the Performance of Innovation-Motivated Acquisitions. In *HBS Corporate Entrepreneurship Research Conference 2005*.

Chesbrough, H. W. (2003). *Open Innovation: The New Imperative for Creating and Profiting from Technology*. Boston, MA: Harvard Business School Press.

Chesbrough, H. W. (2006). *Open Business Models: How to Thrive in the New Innovation Landscape*. Boston, MA: Harvard Business School Press.

Chesbrough, H. W., & Appleyard, M. M. (2007). Open innovation and strategy. *Californian Management Review, 50*(1), 57–76.

Choi, S., & McNamara, G. (2018). Repeating a familiar pattern in a new way: The effect of exploitation and exploration on knowledge leverage behaviors in technology acquisitions. *Strategic Management Journal, 39*(2), 356–378.

Chondrakis, G. (2016). Unique synergies in technology acquisitions. *Research Policy, 45*(9), 1873–1889.

Chondrakis, G., Serrano, C. J., & Ziedonis, R. H. (2020). Information disclosure and the market for acquiring technology companies. *Strategic Management Journal*, forthcoming.

Christensen, C. M., Alton, R., Rising, C., & Waldeck A. (2011). The big idea: the new M&A playbook. *Harvard Business Review, 89*, 48–57.

Clark, H. H., & Brennen, S. E. (1991). Grounding in communication. In: Resnick, L. B., & Levine, J. M. (Eds.), *Perspectives on Socially Shared Cognition* (pp. 222–233), American Psychological Association.

Cloodt, M., Hagedoorn, J., & Van Kranenburg, H. (2006). Mergers and acquisitions: Their effect on the innovative performance of companies in high-tech industries. *Research Policy 35*(5), 642–654.

Coff, R. W. (1999). How Buyers Cope with Uncertainty When Acquiring Firms in Knowledge-Intensive Industries: Caveat Emptor. *Organization Science, 10*(2), 144–161.

Cohen, W. M., & Levinthal, D. A. (1990). Absorptive capacity: A new perspective on learning and innovation. *Administrative Science Quarterly, 35*(1), 128–152.

Colombo, M. G., & Garrone, P. (2006). The impact of M&A on innovation: Empirical results. In B. Cassiman, & M. G. Colombo (Eds.), *Mergers & acquisitions: The innovation impact* (pp. 104–133). Cheltenham, UK and Northampton, MA: Edward Elgar.

Colombo, M. G., & Rabbiosi, L. (2014). Technological similarity, post-acquisition R&D reorganization, and innovation performance in horizontal acquisitions. *Research Policy, 43*(6), 1039–1054.

Colombo, M. G., Piva, E., & Rossi-Lamastra, C. (2014). Open innovation and within-industry diversification in small and medium enterprises: The case of open source software firms. *Research Policy, 43*(5), 891–902.

Comanor, W. S. (1967). Vertical mergers, market power, and the antitrust laws. *American Economic Review, 57*(2), 254–265.

Comino, S., Manenti, F. M., & Parisi, M. L. (2007). From planning to mature: On the success of open source projects. *Research Policy, 36*(10), 1575–1586.

Cosentino, V., Luis, J., & Cabot, J. (2016). Findings from GitHub: Methods, Datasets and Limitations. In *13th Working Conference on Mining Software Repositories* (pp. 137–141), *MSR 2016—Proceedings*. Association for Computing Machinery.

Correia, S. (2017). A feasible estimator for linear models with multi-way fixed effects. Working Paper. Available at: http://scorreia.com/research/hdfe.pdf.

Cox, D. R. (1972). Regression models and life tables. *Journal of the Royal Statistical Society Series B, 34*(2), 187–202.

Coyle, J., & Polsky, G. (2013). Acqui-hiring. *Duke Law Journal, 63*(2), 281–346.

Creswell, J. W., & Plano Clark, V. L. (2011). *Designing and Conducting Mixed Methods Research.* Thousand Oaks, CA: SAGE Publications.

Creswell, J. W. (2014). *Research design. Qualitative, quantitative, and mixed methods approaches.* Thousand Oaks, CA: SAGE Publications.

Crowston, K., Li, Q., Wei, K., Eseryel, U. Y., & Howison, J. (2007). Self-Organization of Teams for Free/Libre Open Source Software Development. *Information and Software Technology, 49* (6), 564–575.

Cunningham, C., Ederer, F., & Ma, S. (2020). Killer Acquisitions. *Journal of Political Economy,* forthcoming.

Cusumano, M., MacCormack, A., Kemerer, C. F., & Crandall, B. (2003). Software Development Worldwide: The State of the Practice. *IEEE Software, 20*(6), 28–34.

Dahlander, L. (2005). Appropriation and appropriability in open source software. *International Journal of Innovation Management, 9*(3), 259–285.

Dahlander, L., & Magnusson, M. G. (2005). Relationships between open source software companies and communities: Observations from Nordic firms. *Research Policy, 34*(4), 481–493.

Dahlander, L., & Gann, D. (2010). How open is innovation? *Research Policy, 39*(6), 699–709.

Dahlander, L., & Magnusson, M. G. (2005). Relationships between open source software companies and communities: Observations from Nordic firms. *Research Policy, 34*(4), 481–493.

Dahlander, L., & Magnusson, M. G. (2008). How do firms make use of open source communities? *Long Range Planning 41*(6), 629–649.

Dahlander, L., & Wallin, M. W. (2006). A man on the inside: Unlocking communities as complementary assets. *Research Policy, 35*(8), 1243–1259.

Dalle, J., den Besten, M., & Menon, C. (2017). Using Crunchbase for economic and managerial research. OECD Science, Technology and Industry Working Papers 2017/08. Available for download: https://www.oecd-ilibrary.org/industry-and-services/using-crunchbase-for-economic-and-managerial-research_6c418d60-en

Dalle, J.-M., & Jullien, N. (2003). 'Libre' Software: Turning Fads into Institutions? *Research Policy, 32*(1), 1–11.

Dam, K. W. (1995). Some economic considerations in the intellectual property protection of software. *The Journal of Legal Studies, 24*(2), 321–377.

Daniel, S. L., Maruping, L., Cataldo, M., & Herbsleb, J. (2018). The Impact of Ideology Fit on Companies and OSS Communities. *Management Information Systems Quarterly, 42*(4), 1069–1096.

Datta, D. K. (1991). Organizational Fit and Acquisition Performance: Effects of Post-Acquisition Integration. *Strategic Management Journal, 12*(4), 281–297.

Datta, D., & Grant, J. (1990). Relationships between type of acquisition, the autonomy given to the acquired firm and acquisition success: an empirical analysis. *Journal of Management, 13*(1), 29–44.

de Man, A.-P., & Duysters, G. (2005). Collaboration and innovation: a review of the effects of mergers, acquisitions and alliances on innovation. *Technovation, 25*(12), 1377–1387.

Dezi, L., Battisti, E., Ferraris, A., Papa, A. (2018). The link between mergers and acquisitions and innovation: A systematic literature review. *Management Research Review, 41*(6), 716–752.

DiBona, C., Ockerbloom, J., & Stone, M. (1999). Introduction. In: C. DiBona, S. Ockman, & M. Stone (Eds.), *Open Sources: Voices of the Open Source Revolution* (pp. 1–17). Sebastopol, CA: O'Reilly & Associates.

Donnelly, W. E. (2009). The Heightened Importance of Thorough Due Diligence in the Current Market Environment. *Practical Compliance & Risk Management for the Securities Industry, 2*(4), 13–18.

Edmondson, A. C., & McManus, S. E. (2007). Methodological Fit in Management Field Research. *Academy of Management Review, 32*(4), 1155–1179.

Eisenhardt, K. M. (1989). Building theories from case study research. *Academy of Management Review, 14*(4), 532–550.

Eisenhardt, K. M., & Graebner, M. E. (2007). Theory Building From Cases: Opportunities and Challenges. *Academy of Management Journal, 50*(1), 25–32.

Ernst, H., & Vitt, J. (2000). The influence of corporate acquisitions on the behaviour of key inventors. *R&D Management, 30*(2), 105–120.

Fang, Y., & Neufeld, D. (2009). Understanding sustained participation in open source software projects. *Journal of Management Information Systems, 25*(4), 9–50.

Faraj, S., von Krogh, G., Monteiro, E., & Lakhani, K. R. (2016). Special Section Introduction—Online Community as Space for Knowledge Flows. *Information Systems Research, 27*(4), 668–684.

Fischer, M., Henkel, J., & Stern, A. D. (2020). Pioneer (Dis-)advantages in Markets for Technology. Harvard Business School Working Paper 19–043. Available at: https://papers.ssrn.com/sol3/papers.cfm?abstract_id=3362959.

Fisher, G. (2019). Online communities and firm advantages. *Academy of Management Review, 44*(2), 279–298.

Fitzgerald, B. (2006). The transformation of Open Source software. *Management Information Systems Quarterly, 30*(3), 587–598.

Foray, D. (2004). *The Economics of Knowledge.* Cambridge, MA: MIT Press.

Fosfuri, A., Giarratana, M. S., & Luzzi, A. (2008). The penguin has entered the building: The commercialization of Open Source Software products. *Organization Science, 19*(2), 292–305.

Franke, N., & Shah, S. (2003). How communities support innovative activities: An exploration of assistance and sharing among end-users. *Research Policy, 32*(1), 157–178.

Franke, N., & von Hippel, E. (2003). Satisfying Heterogeneous User Needs via Innovation Toolkits: The Case of Apache Security Software. *Research Policy, 32*(7), 1199–1215.

Franke, N., Keinz, P., & Klausberger, K. (2013). 'Does this sound like a fair deal?': Antecedents and consequences of fairness expectations in the individual's decision to participate in firm innovation. *Organization Science, 24*(5), 1495–1516.

Fridolfsson, S., & Stennek, J. (2005). Why Mergers Reduce Profits and Raise Share Prices: A Theory of Preemptive Mergers. *Journal of European Economic Association, 3*(5), 1083–1104.

Gallus, J. (2017). Fostering public good contributions with symbolic awards: A large-scale natural field experiment at Wikipedia. *Management Science, 63*(12), 3999–4446

Galpin, T. J., & Herndon, M. (2007). *The Complete Guide to Mergers and Acquisitions.* San Francisco, CA: Jossey-Bass.

Gambardella, A., & von Hippel, E. (2019). Open sourcing as a profit-maximizing strategy for downstream firms. *Strategy Science, 4*(1), 41–57.

Gans, J. S., Hsu, D. H., & Stern, S. (2008). The Impact of Uncertain Intellectual Property Rights on the Market for Ideas: Evidence from Patent Grant Delays. *Management Science, 54*(5), 982–997.

Gentles, S. J., Charles, C., Ploeg, J., & McKibbon, K. A., (2015). Sampling in Qualitative Research: Insights from an Overview of the Methods Literature. *The Qualitative Report, 20*(11), 1772–1789.

Gioia, D. A., Corley, K. G., & Hamilton, A. L. (2013). Seeking Qualitative Rigor in Inductive Research. *Organizational Research Methods 16*(1), 15–31.

Giuri, P., Mariani, M., Brusoni, S., Crespi, G., Francoz, D., Gambardella, A., Garcia-Fontes, W., Geuna, A., Gonzales, R., Harhoff, D., Hoisl, K., Lebas, C., Luzzi, A., Magazzini, L., Nesta, L., Nomaler, Ö., Palomeras, N., Patel, P., Romanelli, M., & Verspagen, B. (2006). Everything you always wanted to know about inventors (but never asked): evidence from the Patval-EU Survey. *CEPR Discussion Paper No. 5752.*

Glaser, B. G., & Strauss, A. L. (1967). *The Discovery of Grounded Theory–Strategies for Qualitative Research.* New Brunswick, NJ: AldineTransaction.

Golden, B. R. (1992). The Past Is the Past–or Is It? The Use of Retrospecitve Accounts as Indicators of Strategy. *Academy of Management Journal, 35*(4), 848–860.

Goldman, R., & Gabriel, R. P. (2005). *Open Source as Business Strategy: Innovation Happens Elsewhere.* San Francisco, CA: Morgan Kaufmann.

Gomes, E., Angwin, D. N., Weber, Y., & Tarba, S. Y. (2013). Critical Success Factors through the Mergers and Acquisitions Process: Revealing Pre- and Post-M&A Connections for Improved Performance. *Thunderbird International Business Review, 55*(1), 13–35.

Goold, M., Campbell, A., & Alexander, M. (1994). *Corporate level strategy: Creating value in the multibusiness company.* New York, NY: Wiley.

Gousios, G. (2013). The GHTorrent dataset and tool suite. In *10th Conference on Mining Software Repositories, MSR 2013—Proceedings* (pp. 233–236). Association for Computing Machinery.

Gousios, G., Vasilescu, B., Serebrenik, A., & Zaidman, A. (2014). Lean GHTorrent: GitHub data on demand. In *11th Conference on Mining Software Repositories* (pp. 384–387)*, MSR 2014–Proceedings.* Association for Computing Machinery.

Gousios, G., & Spinellis, D. (2017). Mining Software Engineering Data from GitHub. In *2017 IEEE/ACM 39th IEEE International Conference on Software Engineering Companion* (pp. 501–502), Buenos Aires, Argentina.

Graebner, M. E. (2004). Momentum and serendipity: How acquired leaders create value in the integration of technology firms. *Strategic Management Journal,* 25(8-9), 751–777.

Graebner, M. E., & Eisenhardt, K. M. (2004). The Seller's Side of the Story: Acquisition as Courtship and Governance as Syndicate in Entrepreneurial Firms. *Administrative Science Quarterly, 49*(3), 366–403.

Graebner, M. E., Eisenhardt, K. M., & Roundy, P. T. (2010). Success and Failure in Technology Acquisitions: Lessons for Buyers and Sellers. *Academy of Management Perspectives, 24*(3), 73–92.

Grand, S., von Krogh, G., Leonard, D., & Swap, W. (2004). Resource Allocation Beyond Firm Boundaries: A Multi-Level Model for Open Source Innovation. *Long Range Planning, 37*(6), 591–610.

Granstrand, O. (1999). *The Economics and Management of Intellectural Property.* Cheltenham, UK: Edward Elgar.

Granstrand, O., Bohlin, E., Oskarsson, C., & Sjöberg, N. (1992). External Technology Acquisition in Large Multi-Technology Corporations. *R&D Management, 22*(2), 111–134.

Grimpe, C., & Hussinger K. (2008). Pre-empting technology competition through firm acquisitions. *Economics Letters, 100*(2), 189–191.

Gruber, M., & Henkel, J. (2006). New Ventures Based on Open Innovation—An Empirical Analysis of Start-Up Firms in Embedded Linux. *International Journal of Technology Management, 33*(4), 356–372.

Haleblian, J., & Finkelstein, S. (1999). The influence of organizational acquisition experience on acquisition performance: A behavioral learning perspective. *Administrative Science Quarterly, 44*(1), 29–56.

Hambrick, D. C, & Cannella, A. A. (1993). Relative standing: A framework for understanding departures of acquired executives. *Academy of Management Journal, 36*(4), 733–762.

Hann, I., Roberts, J. A., Slaughter, S. A., & Fielding, R. T. (2002). Economic Incentives for Participating in Open Source Software Projects. *Proceedings of the International Conference on Information Systems,* ICIS 2002, Barcelona, Spain.

Hann, I., Roberts, J. A., & Slaughter, S. A. (2013). All Are Not Equal: An Examination of the Economic Returns to Different Forms of Participation in Open Source Software Communities. *Information Systems Research, 24*(3), 520–538.

Hansen, R. G. (1987). A theory for the choice of exchange medium in mergers and acquisitions. *Journal of Business, 60*(1), 75–95.

Harhoff, D., Henkel, J., & von Hippel, E. (2003). Profiting from voluntary information spillovers: how users benefit by freely revealing their innovations. *Research Policy, 32*(10), 1753–1769.

Hars, A., & Ou, S. (2002). Working for Free? Motivations for Participating in Open-Source Projects. *International Journal of Electronic Commerce, 6*(3), 25–39.

Haspeslagh, P. C., & Jemison, D. B. (1991), *Managing Acquisitions: Creating Value Through Corporate Renewal.* New York, NY: The Free Press.

Haunschild, P. R., (1994). How Much Is That Company Worth? Interorganizational Relationships, Uncertainty, and Acquisition Premiums. *Administrative Science Quarterly, 39*(3), 391–411.

Hayward, M. L. A. (2002). When do firms learn from their acquisition experience? Evidence from 1990 to 1995. *Strategic Management Journal, 23*(1), 21–39.

Hayward, M. L. A., & Hambrick, D. (1997). Explaining the Premiums Paid for Large Acquisitions: Evidence of CEO Hubris. *Administrative Science Quarterly, 42*(1), 103–127.

He, V. F., Puranam, P., Shrestha, Y. R., & von Krogh, G. (2020). Resolving governance disputes in communities: A study of software license decisions. *Strategic Management Journal, 41*(10), 1837–1868

Hecker, F. (1999). Setting Up Shop: The Business of Open-Source Software. *IEEE Software, 16*(1), 45–51.

Henkel, J. (2004). Open Source Software from Commercial Firms—Tools, Complements, and Collective Invention. *ZfB-Ergänzungsheft, 74*(4).

Henkel, J. (2006). Selective revealing in open innovation processes: The case of embedded Linux. *Research Policy, 35*(7), 953–969.

Henkel, J. (2009). Champions of revealing—The role of open source developers in commercial firms. *Industrial and Corporate Change, 18*(3), 435–471.

Henkel, J., Baldwin, C. Y., & Shih, W. (2013). IP modularity: Profiting from innovation by aligning product architecture with intellectual property. *California Management Review 55*(4), 65–82.

Henkel, J., Schöberl, S., & Alexy, O. (2014). The emergence of openness: How and why firms adopt selective revealing in open innovation. *Research Policy, 43*(5), 879–890.

Henkel, J., & von Hippel E. (2005). Welfare implications of user innovation. *Journal of Technology Transfer, 30*(1–2), 73–87.

Hepp, D. A. (2017). *When Young Firms Change Their Innovation Strategy—The Dynamics of Openness*. Munich, Germany: Verlag Dr. Hut.

Hernandez, E., & Menon, A. (2018). Acquisitions, node collapse, and network revolution. *Management Science, 64*(4), 1652–1671.

Hernandez, E., & Shaver, J. M. (2019). Network synergy. *Administrative Science Quarterly, 64*(1), 171–202.

Hertel, G., Niedner, S., & Herrmann, S. (2003). Motivation of software developers in open source projects: An internet-based survey of contributors to the Linux kernel. *Research Policy, 32*(7), 1159–1177.

Himanen, P. (2001). *The Hacker Ethic and the Spirit of the Information Age*. New York, NY: Random House.

Hitt, M. A., Hoskisson, R. E., & Ireland, R. D. (1990). Mergers and acquisitions and management commitment to innovation in M-form firms. *Strategic Management Journal, 11*(5), 29–47.

Hitt, M. A., Hoskisson, R. E., Ireland, R. D., & Harrison, J. S. (1991). Effects of acquisitions on R&D inputs and outputs. *Academy ofManagement Journal, 34*(3), 639–706.

Hitt, M. A., Hoskisson, R. E., Johnson, R. A., & Moesel, D. D. (1996). The market for corporate control and firm innovation. *Academy of Management Journal, 39*(5), 1084–1119.

Hlavka, T. B. (2019). *Acquisitions in the Information and Communication Technology Industry—Buyer Capabilities, Technology Hype, and Target Maturity*. Berlin, Germany: Neopubli.

Ho, S. Y., & Rai, A. (2017). Continued voluntary participation intention in firm-participating open source software projects. *Information Systems Research, 28*(3), 603–625.

Homburg, C., & Bucerius, M. (2005). A marketing perspective on mergers and acquisitions: how marketing integration affects postmerger performance. *Journal of Marketing, 69*(1), 95–113.

Homburg, C., & Bucerius, M. (2006). Is speed of integration really a success factor of mergers and acquisitions? An analysis of the role of internal and external relatedness. *Strategic Management Journal 27*(4), 347–367.

Homburg, C., Hahn, A., Bornemann, T., & Sandner, P. G. (2014). The Role of Chief Marketing Officers for Venture Capital Funding: Endowing New Ventures with Marketing Legitimacy. *Journal of Marketing Research, 51*(5), 625–644.

Huber, G. P., & Power, D. J. (1985). Research Notes and Communications—Retrospective Reports of Strategic-Level Managers: Guidelines for Increasing Their Accuracy. *Strategic Management Journal, 6*(2), 171–180.

Hussinger, K. (2010). On the importance of technological relatedness: SMEs versus large acquisition targets. *Technovation, 30*(1), 57–64.

Iacus, S., King, G., & Porro, G. (2012). Causal inference without balance checking: Coarsened exact matching. *Political Analysis, 20*(1), 1–24.

Jensen, M. C. (1986). Agency costs of free cash flow, corporate finance, and takeovers. The *American Economic Review, 76*(2), 323–329.

Jeppesen, L. B., & Frederiksen, L. (2006). Why do users contribute to firm-hosted user communities? The case of computer-controlled music instruments. *Organization Science, 17*(1), 45–63.

Jemison, D., & Sitkin, S. (1986). Corporate Acquisitions: A Process Perspective. *Academy of Management Review, 11*(1), 145–163.

Jiang, Q., Tan, C.-H., Sia, C. L., & Wei, K. K. (2019). Followership in an Open-Source Software Project and its Significance in Code Reuse. *Management Information Systems Quarterly, 43*(4), 1303–1319.

Jick, T. D. (1979). Mixing Qualitative and Quantitative Methods: Triangulation in Action. *Administrative Science Quarterly, 24*(4), 602–611.

Jones, C. (2003). Variations in Software Development Practices. *IEEE Software, 20*(6), 22–27.

Kaal, W. A. (2016). Private Fund Investor Due Diligence: Evidence from 1995–2015. *Review of Banking and Financial Law, 36*, 257–313.

Kalliamvakou, E., Gousios, G., Blincoe, K., Singer, L., German, D. M., & Damian, D. (2016). An in-depth study of the promises and perils of mining GitHub. *Empirical Software Engineering, 21*(5), 2035–2071.

Kane, G. C., & Ransbotham, S. (2016). Content as community regulator: The recursive relationship between consumption and contribution in open collaboration communities. *Organization Science, 27*(5), 1258–1274.

Kapoor, R., & Lim, K. (2007). The impact of acquisitions on the productivity of inventors at semiconductor firms: A synthesis of knowledge-based and incentive-based perspectives. *Academy of Management Journal, 50*(5), 1133–1155.

Karim, S. (2006). Modularity in organizational structure: The reconfiguration of internally developed and acquired business units. *Strategic Management Journal, 27*(9), 799–823.

Karim, S., & Capron, L. (2016). Reconfiguration: Adding, redeploying, recombining and divesting resources and business units. *Strategic Management Journal, 37*(13), 799–823.

Karim, S., & Mitchell, W. (2000). Path-dependent and path-breaking change: reconfiguring business resources following acquisitions in the U.S. medical sector, 1978–1995. *Strategic Management Journal, 21*(10–11), 1061–1081.

Kato, J., & Schoenberg, R. (2014). The impact of post-merger integration on the customer–supplier relationship. *Industrial Marketing Management, 43*(2), 335–345.

Kaul, A., & Wu, B. (2015). A capabilities-based perspective on target selection in acquisitions: A Capabilities-Based Perspective on Target Selection. *Strategic Management Journal 37*(7), 1220–1239.

Kavusan, K., Ates, N. Y., & Nadolska, A. (2020). Acquisition target selection and technological relatedness: The moderating role of Top Management Team demographic faultlines. *Strategic Organization 29*(1), 1–29.

Kengelbach, J., Keienburg, G., Schmid, T., Sievers, S., Gjerstad, K., Nielsen, J., & Decker, W. (2017). *The 2017 M&A Report: The Technology Takeover.* Available at: https://www.bcg.com/publications/2017/corporate-development-finance-technology-digital-2017-m-and-a-report-technology-takeover.

Kim, J. D. (2019). Predictable exodus: Startup acquisitions and employee departures. *SSRN Electronic Journal.* Available at: https://ssrn.com/abstract=3252784

King, D. R., Dalton, D. R., Daily, C. M., & Covin, J. G. (2004). Meta-analyses of post-acquisition performance: Indications of unidentified moderators. *Strategic Management Journal, 25*(2), 187–200.

Klapper, H., & Reitzig, M. (2018). On the effects of authority on peer motivation: Learning from Wikipedia, *Strategic Management Journal, 39*(8), 2178–2203.

Koeplin, J., Sarin, A., Shapiro, A. C. (2000). The private company discount. *Journal of Applied Corporate Finance, 12*(4), 48–55.

Krebs, D. L. (1970). Altruism—Examination of Concept and a Review of Literature. *Psychological Bulletin, 73*(4), 258–302.

Krug, J. A., & Nigh, D. (2001). Executive perceptions in foreign and domestic acquisitions: An analysis of foreign ownership and its effect on executive fate. *Journal of World Business, 36*(1), 85–98.

Kurokawa, S. (1997). Make-or-Buy Decisions in R&D: Small Technology Based Firms in the United States and Japan. *IEEE Transactions on Engineering Management, 44*(2), 124–134.

Kuzel, A. J. (1992). Sampling in qualitative inquiry. In B. F. Crabtree, & W. L. Miller (Eds.), *Doing qualitative research* (pp. 31–44). Thousand Oaks, CA: SAGE Publications.

Lakhani, K. R., & von Hippel, E. (2003). How Open Source Software Works: 'Free' User-to-User Assistance. *Research Policy, 32*(6), 923–43.

Lakhani, K. R., & Wolf, R. G. (2005). Why hackers do what they do: Understanding motivation and effort in free/open source software projects. In J. Feller, B. Fitzgerald, S. Hissam, & K. R. Lakhani (Eds.), *Perspectives on Free and Open Source Software* (pp. 3–22). Cambridge, MA: MIT Press.

Lambe, C. J., & Spekman, R. E. (1997). Alliances, external technology acquisition, and discontinuous technological change. *Journal of Product Innovation Management, 14*(2), 102–116.

Larsson, R., & Finkelstein, S. (1999). Integrating strategic, organizational, and human resource perspectives on mergers and acquisitions: A case survey of synergy realization. *Organization Science, 10*(1), 1–26.

Lee, G. K., & Cole, R. E. (2003). From a Firm-Based to a Community-Based Model of Knowledge Creation: The Case of the Linux Kernel Development. *Organization Science, 14*(6), 633–649.

Lerner, J., & Tirole, J. (2002). Some simple economics of open source. *The Journal of Industrial Economics, 50*(2), 197–234.

Lerner, J., & Tirole, J. (2005). The scope of open source licensing. *Journal of Law, Economics, and Organization, 21*(1), 20–56.

Liao, Z., Zhao, B., Liu, S., Jin, H., He, D., Yang, L., Wu, J., & Zhang, Y. (2019). A Prediction Model of the Project Life-Span in Open Source Software Ecosystem. *Mobile Networks and Applications, 24*, 1382–1391.

Lin, Y.-W. (2006). Hybrid innovation: The dynamics of collaboration between the FLOSS community and corporations. *Knowledge, Technology & Policy, 18*(4), 86–100.

MacCormack, A. D., & Verganti, R. (2003). Managing the Sources of Uncertainty: Matching Process and Context in Software Development. *Journal of Product Innovation Management, 20*(3), 217–232.

MacCormack, A. D., Rusnak, J., & Baldwin, C. Y. (2006). Exploring the Structure of Complex Software Designs: An Empirical Study of Open Source and Proprietary Code. *Management Science, 52*(7), 1015–1030.

Machlup, F., & Penrose E. (1950). The patent controversy in the nineteenth century. *The Journal of Economic History, 10*(1), 1–29.

MacMillan, I. C., & McGrath, R. G. (2002). Crafting R&D Project Portfolios. *Research-Technology Management, 45*(5), 48–59.

Macredie, R. D., & Mijinyawa, K. (2011). A theory-grounded framework of Open Source Software adoption in SMEs. *European Journal of Information Systems, 20*, 237–250.

Madanmohan, T. R., & De', R. (2004). Open Source Reuse in Commercial Firms. *IEEE Software, 21*(6), 62–69.

Makri, M., Hitt, M. A., & Lane, P. J. (2010), Complementary technologies, knowledge relatedness, and invention outcomes in high technology mergers and acquisitions. *Strategic Management Journal, 31*(6), 602–628.

Mahoney, J., & Goertz, G. (2006). A tale of two cultures: Contrasting quantitative and qualitative research. *Political Analysis, 14*(3), 227–249.

Manne, H. (1965). Mergers and the Market for Corporate Control. *Journal of Political Economy, 73*(2), 110–120.

Markus, M. L., Manville, B., & Agres, C. E. (2000). What Makes a Virtual Organization Work? *Sloan Management Review, 42*(1), 13–26.

Mayring, P. (2014). *Qualitative Content Analysis. Theoretical Foundation, Basic Procedures, and Software Solution.* gesis—Leibniz-Institut für Sozialwissenschaften, Klagenfurth. Available at: https://nbn-resolving.org/urn:nbn:de:0168-ssoar-395173.

McCarthy, K., & Aalbers, H. (2016). Technological acquisitions: The impact of geography on post-acquisition innovative performance. *Research Policy, 45*(9), 1818–1832.

Meckl, R. (2004). Organising and Leading M&A Projects. *International Journal of Project Management, 22*(6), 455–462.

Meglio, O. (2009). Measuring Performance in Technology-Driven M&AS: Insights from a Literature Review. In: C.L. Cooper, & S. Finkelstein (Eds.), *Advances in Mergers and Acquisitions Vol. 8* (pp. 103–118), Bingley, UK: Emerald Group Publishing Limited.

Mehra, A., Dewan, R., & Freimer, M. (2011). Firms as incubators of opensource software. *Information Systems Research, 22*(1), 22–38.

Merriam, S. B. (2009). *Qualitative Research—A Guide to Design and Implementation.* San Francisco, CA: Jossey-Bass.

Miles, M. B., & Huberman, A. M. (1994*). Qualitative data analysis: An expanded sourcebook (2nd ed.).* Newbury Park, CA: SAGE Publications.

Mintzberg, H., Raisinghani, D., & Theoret, A. (1976). The Structure of "Un-structured" Decision Processes. *Administrative Science Quarterly, 21*(2), 246–275.

Mockus, A., Fielding, R. T., & Herbsleb, J. D. (2005). Two Case Studies of Open Source Software Development: Apache and Mozilla. In: J. Feller, B. Fitzgerald, S. A. Hissam, & K. R. Lakhani (Eds.), *Perspectives on Free and Open Source Software* (pp. 163–209). Cambridge, MA: MIT Press.

Moeller, S. B., Schlingemann, F. P., & Stulz, R. M. (2005). Wealth destruction on a massive scale? A study of acquiring firm returns in the recent merger wave. *The Journal of Finance, 60*(2), 757–782.

Moody, G. (2001). Rebel Code—Inside Linux and the Open Source Revolution. Cambridge, MA: Perseus Publishing.

Morgan, L., & Finnegan, P. (2007). Benefits and Drawbacks of Open Source Software: An Exploratory Study of Secondary Software Firms. In: J. Feller, B. Fitzgerald, W. Scacchi, &

A. Sillitti (Eds.), *Open Source Development, Adoption and Innovation. OSS 2007. IFIP — The International Federation for Information Processing* (vol. 234; pp. 307–312). Boston, MA: Springer.

Morrison, P. D., Roberts, J. H., & von Hippel, E. (2000). Determinants of User Innovation and Innovation Sharing in a Local Market. *Management Science, 46*(12), 1513–1527.

Mustonen, M. (2003). Copyleft—the economics of Linux and other open source software. *Information Economics and Policy, 15*(1), 99–121.

Nagaraj, A., & Piezunka, H. (2020). How Competition Affects Contributions to Open Source Platforms: Evidence from OpenStreetMap and Google Maps. Working Paper, University of California, Berkeley, CA. Available at: http://abhishekn.com/files/openstreetmap_google_feb2020.pdf.

Nagle, F. (2018a). Learning by contributing: Gaining competitive advantage through contribution to crowdsourced public goods. *Organization Science, 29*(4), 569–587.

Nagle, F. (2018b). Open source software and firm productivity. *Management Science, 65*(3), 1191–1215.

Nagle, F. (2019). Government technology policy, social value, and national competitiveness. Harvard Business School Strategy Unit Working Paper 19–103. Available at: https://ssrn.com/abstract=3355486

Nahavandi, A., & Malekzadeh, A. (1988). Acculturation in acquisitions and acquisitions. *Academy of Management Review, 13*(1), 79–90.

Nielsen, B. B., & Gudergan, S. (2012). Exploration and exploitation fit and performance in international strategic alliances. *International Business Review, 21*(4), 558–574.

Oh, W., & Jeon, S. (2007). Membership herding and network stability in the open source community: The ising perspective. *Management Science, 53*(7), 1086–1101.

O'Mahony, S. (2003). Guarding the commons: how community managed software projects protect their work. *Research Policy, 32*(7), 1179–1198.

O'Mahony, S., & Bechky, B. A. (2008). Boundary organizations: Enabling collaboration among unexpected allies. *Administrative Science Quarterly, 53*(3), 422–459.

O'Mahony, S., & Ferraro, F. (2007). The emergence of governance in an open source community. *Academy of Management Journal, 50*(5), 1079–1106.

O'Mahony, S., & Lakhani, K. R. (2011). Organizations in the shadow of communities. In C. Marquis, M. Lounsbury, & R. Greenwood (Eds.), *Communities and organizations* (pp. 3–36). Bingley, UK: Emerald Group Publishing Limited.

Olie, R. (1994). Shades of culture and institution in international acquisitions. *Organization Studies, 15*(3), 381–401.

Ouimet, P., & Zarutskie, R. (2012). Acquiring Labor, Working Papers 11–32, Center for Economic Studies, U.S. Census Bureau, USA. Available at: https://papers.ssrn.com/sol3/papers.cfm?abstract_id=1571891

Oxley, J. E., Rivkin, J. W., & Ryall, M. D. (2010). The strategy research initiative: Recognizing and encouraging high-quality research in strategy. *Strategic Organization, 8*(4), 377–386.

Pablo, A. L. (1996). Acquisition Decision-Making Processes: The Central Role of Risk. *Journal of Management, 22*(5), 723–746.

Papamichail, M., Diamantopoulos, T., & Symeonidis, A. (2016). User-perceived source code quality estimationbased on static analysis metrics. In *2016 IEEE International Conference on Software Quality, Reliability and Security* (pp. 100–107), Vienna, Austria.

Paruchuri, S., Nerkar, A., & Hambrick, D. C. (2006). Acquisition integration and productivity losses in the technical core: Disruption of inventors in acquired companies. *Organization Science, 17*(5), 545–562.

Patton, M. Q. (1990). *Qualitative evaluation and research methods.* Newbury Park, CA: SAGE Publications.

Pennings, J. M., Barkema, H., & Douma, S. (1994). Organizational Learning and Diversification. *Academy of Management Journal, 37*(3), 608–640.

Perens, B. (1999). *The Open Source Definition.* In: C. DiBona, S. Ockman, & M. Stone (Eds.), *Open Sources: Voices of the Open Source Revolution* (pp. 171–189). Sebastopol, CA: O'Reilly & Associates.

Perr, J., Appleyard, M. M., & Sullivan, P. (2010). Open for business: emerging business models in open source software. *International Journal of Technology Management, 52*(3/4), 432–456.

Pettigrew, A. M. (1990). Longitudinal field research on change: Theory and practice. *Organization Science, 1*(3), 267–292.

Pich, M. T., Loch, C. H., & De Meyer, A. (2002). On Uncertainty, Ambiguity, and Complexity in Project Management. *Management Science, 48(8), 1008–1023.*

Plant, A. (1934). The Economic Theory Concerning Patents for Inventions. *Economica, 1*(1), 30–51.

Prabhu, J. C., Chandy, R. K., & Ellis, M. E. (2005). The Impact of Acquisitions on Innovation: Poison pill, placebo, or tonic? *Journal of Marketing, 69*(1), 114–130.

Puranam, P., Singh, H., & Chaudhuri, S. (2009). Integrating acquired capabilities: When structural integration is (un)necessary. *Organization Science, 20*(2), 313–328.

Puranam, P., Singh, H., & Zollo, M. (2003). A bird in the hand or two in the bush?: Integration trade-offs in technology-grafting acquisi tions. *European Management Journal, 21*(2), 179–184.

Puranam, P., Singh, H., & Zollo, M. (2006). Organizing for innovation: Managing the coordination-autonomy dilemma in technology acquisitions. *Academy of Management Journal, 49*(2), 263–280.

Puranam, P., & Srikanth, K. (2007). What they know vs. what they do: how acquirers leverage technology acquisitions. *Strategic Management Journal, 28*(8), 805–825.

Rabier, M. R. (2017). Acquisition Motives and the Distribution of Acquisition Performance. *Strategic Management Journal, 38*(13), 2666–2681.

Ranft, A. L. (2006). Knowledge preservation and transfer during post-acquisition integration. In: C. L. Cooper, & S. Finkelstein (Eds.), *Advances in Mergers and Acquisitions Vol. 5* (pp. 51–67). Bingley, UK: Emerald Group Publishing Limited.

Ranft, A. L., & Lord, M. D. (2002). Acquiring new technologies and capabilities: a grounded model of acquisition implementation. *Organization Science, 13*(4), 420–41.

Ransbotham, S., & Mitra, S. (2010). Target Age and the Acquisition of Innovation in High-Technology Industries. *Management Science, 56*(11), 2076–2093.

Rao, V. R., Yu, Y., & Umashankar, N. (2016). Anticipated vs. actual synergy in merger partner selection and postmerger innovation. *Marketing Science, 35*(6), 934–952.

Raymond, E. S. (2001). *The Cathedral and the Bazaar (2nd ed.).* Sebastopol, CA: O'Reilly & Associates.

Ren, Y., Kraut, R., & Kiesler, S. (2007). Applying common identity and bond theory to design of online communities. *Organization Studies, 28*(3), 377–408.

Ren, Y., Harper, F. M., Drenner, S., Terveen, L., Kiesler, S., Riedl, J., & Kraut, R. (2012). Building member attachment in online communities: Applying theories of group identity and interpersonal bonds. *Management Information Systems Quarterly, 36*(3), 841–864.

Reuer, J., & Ragozzino, R. (2008). Adverse selection and M&A design: The roles of alliances and IPOs. *Journal of Economic Behavior and Organization, 66*(2), 195–212.

Rogan, M. (2014). Too Close for Comfort? The Effect of Embeddedness and Competitive Overlap on Client Relationship Retention Following an Acquisition. *Organization Science, 25*(1), 185–203.

Rogan, M., & Greve, H. R. (2015). Resource dependence dynamics: Partner reactions to mergers. *Organization Science, 26*(1), 239–255.

Rogan, M., & Sorenson, O. (2014). Picking a (Poor) Partner: A Relational Perspective on Acquisitions. *Administrative Science Quarterly, 59*(2), 301–329.

Rolandsson, B., Bergquist, M., & Ljungberg, J. (2011). Open source in the firm: Opening up professional practices of software development. *Research Policy, 40*(4), 576–587.

Roll, R. (1986). The Hubris Hypothesis of Corporate Takeovers. *Journal of Business, 59*(2), 197–216.

Ryan, R. M., & Deci, E. L. (2000). Intrinsic and Extrinsic Motivations: Classic Definitions and New Directions. *Contemporary Educational Psychology, 25*(1), 54–67.

Salter, M. S., & Weinhold, W. A. (1981). Choosing compatible acquisitions. *Harvard Business Review, 59*(1).117–127.

Sanatinia, A., & Noubir, G. (2016). On GitHub's Programming Languages. *CoRR*. Available at: https://arxiv.org/abs/1603.00431

Scacchi, W. (2004). Free and Open Source Development Practices in the Game Community. *IEEE Software, 21*(1), 59–66.

Schaarschmidt, M., Walsh, G., & von Kortzfleisch, H. F. (2015). How Do Firms Influence Open Source Software Communities? A Framework and Empirical Analysis of Different Governance Modes. *Information and Organization, 25*(2), 99–114.

Schweiger, D. M., & DeNisi, A. S. (1991). Communication with employees following an acquisition: A longitudinal field experiment. *Academy of Management Journal, 34*(1), 110–135.

Schweiger, D. M., & Weber, Y. (1989). Strategies for managing human resources during mergers and acquisitions: An empirical investigation. *Human Resource Planning, 12*(2), 69–86.

Schwenk, C. R. (1984). Cognitive Simplification Processes in Strategic Decision-Making. *Strategic Management Journal, 5*(2), 111–128.

Seidler, J. (1974). On using informants: A technique for collecting quantitative data and controlling measurement error in organization analysis. *American Sociological Review, 39*(6), 816–831.

Seo, E., Nagle, F., & Shah, S. K. (2020). A Little Help from My Friends: How Receiving Assistance Affects Participation in Online Knowledge-Sharing Communities. Harvard Business School Strategy Unit Working Paper No. 21–026, Available at SSRN: https://ssrn.com/abstract=3680076.

Senyard, A., & Michlmayr, M. (2004). *How to Have a Successful Free Software Project.* Paper presented at the 11th Asia-Pacific Software Engineering Conference, Busan, South Korea.

Seth, A. (1990). Value creation in acquisitions: A re-examination of performance issues. *Strategic Management Journal, 11*(2), 99–115.

Shah, S. K. (2006). Motivation, governance & the viability of hybrid forms in open source software development. *Management Science, 52*(7), 1000–1014.

Shah, S. K., & Nagle, F. (2020). Why Do User Communities Matter for Strategy? *Strategic Management Review,* forthcoming.

Sharma, D. S., & Ho, J. (2002). The impact of acquisitions on operating performance: Some Australian evidence. *Journal of Business Finance and Accounting, 29*(1), 155–200.

Shen, J.-C., & Reuer, J. J. (2005). Adverse Selection in Acquisitions of Small Manufacturing Firms: A Comparison of Private and Public Targets. *Small Business Economics 24*(4), 393–407.

Shrivastava, P. (1986). Postmerger integration. *Journal of Business Strategy, 7*(1), 65–76.

Shriver, S. K., Nair, H. S., & Hofstetter, R. (2013). Social ties and user-generated content: Evidence from an online social network. *Management Science, 59*(6), 1425–1443.

Siggelkow, N. (2007). Persuasion With Case Studies. *Academy of Management Journal, 50*(1), 20–24.

Sims, J., & Woodard, C. J. (2020). Community interactions at crowd scale: hybrid crowds on the GitHub platform. *Innovation, 22*(2), 105–127.

Singh, H., & Montgomery, C. A. (1987). Corporate acquisition strategies and economic performance. *Strategic Management Journal, 8*(4), 377–386.

Snow, C. C., & Thomas, J. B. (1994). Field research methods in strategic management: Contributions to theory building and testing. *Journal of Management Studies, 31*(4), 457–480.

Sojer, M. (2010). *Reusing Open Source Code: Value Creation and Value Appropriation Perspectives on Knowledge Reuse.* Wiesbaden, Germany: Gabler Verlag.

Sorenson, O., & Stuart, T. E. (2008). Bringing the Context Back In: Settings and the Search for Syndicate Partners in Venture Capital Investment Networks. *Administrative Science Quarterly, 53*(2), 266–294.

Spaeth, S., von Krogh, G., & He, F. (2015). Research Note — Perceived firm attributes and intrinsic motivation in sponsored open source software projects. *Information Systems Research, 26*(1), 224–237.

Stahl, G. K., & Voigt, A. (2008). Do cultural differences matter in mergers and acquisitions? A tentative model and examination. *Organization Science, 19*(1), 160–176.

Stallman, R. (1984). *The GNU Manifesto.* Available at: http://www.gnu.org/gnu/manifesto.html.

Steensma, H. K., & Corley, K. G. (2000). On the Performance of Technology-Sourcing Partnerships: The Interaction Between Partner Interdependence and Technology Attributes. *Academy of Management Journal, 43*(6), 1045–1067.

Stein, J. P. (2017). *Technology-focused Acquisitions: Performance and Functionality as Differentiators.* Oldenbourg, Germany: De Gruyter.

Stewart, K., Ammeter, A. P., & Maruping, L. M. (2006). Impacts of License Choice and Organizational Sponsorship on User Interest and Development Activity in Open Source Software Projects. *Information Systems Research, 17*(2), 126–145.

Stewart, K., & Gosain, S. (2006). The Impact of Ideology on Effectiveness in Open Source Software Development Teams, *Management Information Systems Quarterly, 30*(2), 291–314.

Strauss, A. L. (1987). *Qualitative Analysis for Social Scientists*. Cambridge, UK: Cambridge University Press.

Strauss, A. L., & Corbin, J. M. (1998). *Basics of Qualitative Research—Techniques and Procedures for Developing Grounded Theory 2*. Thousand Oaks, CA: SAGE Publications.

Sudarsanam, S. (2003). *Creating Value from Mergers and Acquisitions*. Harlow, UK: Pearson Education.

Subramaniam, C., Sen, R., & Nelson, M. L. (2009). Determinants of open source software project success: A longitudinal study. *Decision Support Systems, 46*(2), 576–585.

Swaminathan, V., Murshed, F., & Hulland, J. (2008). Value creation following merger and acquisition announcements: the role of strategic emphasis alignment. *Journal of Marketing Research, 45*(1), 33–47.

Tanriverdi, H., & Venkatraman, N. (2005). Knowledge relatedness and the performance of multibusiness firms. *Strategic Management Journal, 26*(2), 97–119.

Teigland, R., Di Gangi, P. M., Flålten, B.-T., Giovacchini, E., & Pastorino, N. (2014). Balancing on a tightrope: Managing the boundaries of a firm-sponsored OSS community and its impact on innovation and absorptive capacity. *Information and Organization, 24*(1), 25–47.

Ter Wal, A. L. J., Alexy, O., Block, J., & Sandner, P. G. (2016). The Best of Both Worlds: The Benefits of Open-specialized and Closed-diverse Syndication Networks for New Ventures' Success. *Administrative Science Quarterly, 61*(3), 393–432.

Thomson, R., Dettmar, S., & Garay, M. (2018). The state of the deal: M&A trends 2018. Available at: https://www2.deloitte.com/content/dam/Deloitte/us/Documents/mergers-acqisitions/us-mergers-acquisitions-2018-trends-report.pdf.

Trautwein, F. (1990). Merger motives and merger prescriptions. *Strategic Management Journal, 11*(4), 283–295.

Tsay, J., Dabbish, L., & Herbsleb, J. (2014). Influence of social and technical factors for evaluating contribution in GitHub. *Proceedings of the 36th international conference on Software engineering*, Hyderabad, India.

Tuomi, I. (2002). *Networks of Innovation*. Oxford, UK: Oxford University Press.

Valentini, G. (2012). Measuring the effect of M&A on patenting quantity and quality. *Strategic Management Journal, 33*(3), 336–346.

Valentini, G. (2016). The impact of M&A on rivals' innovation strategy. *Long Range Planning, 49*(2), 241–249.

Vasilescu, B., van Schuylenburg, S., Wulms, J., Serebrenik, A., & van den Brand, M. G. J. Continuous Integration in a Social-Coding World: Empirical Evidence from GitHub. *Proceedings of the 2014 IEEE International Conference on Software Maintenance and Evolution* (pp. 401–405), Victoria, BC.

Very, P., Lubatkin, M., Calori, R., & Veiga, J. (1997). Relative standing and the performance of recently acquired European firms. *Strategic Management Journal, 18*(8), 593–614.

Veugelers, R. (2006). Literature review on M&A and R&D. In: B. Cassiman, & M. G. Colombo (Eds.), *Merger and Acquisitions—The Innovation Impact* (pp. 37–62). Cheltenham, UK: Edward Elgar Publishing.

Viseur, R. (2012). Forks impacts and motivations in free and open source projects. *International Journal of Advanced Computer Science and Applications, 3*(2), 117–122.

Vixie, P. (1999). Software Engineering. In: C. DiBona, S. Ockman, & M. Stone (Eds.), *Open Sources: Voices of the Open Source Revolution* (pp. 91–101). Sebastopol, CA: O'Reilly & Associates.

von Hippel, E. (2001). Innovation by User Communities: Learning from Open-Source Software. *MIT Sloan Management Review, 42*(4), 82–86.

von Krogh, G. F., Spaeth, S., & Lakhani, K. R. (2003). Community, joining, and specialization in open source software innovation: A case study. *Research Policy, 32*(7), 1217–1241.

von Hippel, E., & von Krogh, G. (2003). Open source software and the "private-collective" innovation model: issues for organization science. *Organization Science, 14*(2), 209–223.

von Krogh, G., Haefliger, S., Spaeth, S., & Wallin, M. W. (2012). Carrots and rainbows: Motivation and social practice in open source software development. *Management Information Systems Quarterly, 36*(2), 649–676.

von Krogh, G., & von Hippel, E. (2006). The promise of research on open source software. *Management Science, 52*(7), 975–983.

Walter, G. A., & Barney, J. B. (1990). Research notes and communications management objectives in mergers and acquisitions. *Strategic Management Journal, 11*(1), 79–86.

Wang, S., & Tambe, P. (2020). Star Developers and Open Source Software. Working paper. Mack institute for Innovation Management, University of Pennsylvania, PA. Available at: https://mackinstitute.wharton.upenn.edu/2020/star-developers-and-open-source-software/.

Warner, A. G. (2003). Buying versus building competence: acquisition patterns in the information and telecommunications industry 1995–2000. *International Journal of Innovation Management, 7*(4), 395–415.

Warner, A. G., Fairbank, J. F., & Steensma, H. K. (2006). Managing Uncertainty in a Formal Standards-Based Industry: A Real Options Perspective on Acquisition Timing. *Journal of Management 32*(2), 279–298.

Weber, S. (2004). *The Success of Open Source.* Cambridge, MA: Harvard University Press.

Weber, Y., Tarba, S., & Öberg, C. (2013). *A Comprehensive Guide to Mergers & Acquisitions– Managing the Critical Success Factors Across Every Stage of the M&A Process.* Upper Saddle River, NJ: FT Press.

Weber, Y., & Schweiger, D. (1992). Top management culture in mergers and acquisitions: A lesson in anthropology. *International Journal of Conflict Management, 3*(4), 285–302.

Wernerfelt, B. (1984). A Resource-Based View of the Firm. *Strategic Management Journal, 5*(2), 171–180.

West, J. (2003). How open is open enough? *Research Policy, 32*(7), 1259–1285.

West, J., & Gallagher, S. (2006). Challenges of Open Innovation: The Paradox of Firm Investment in Open Source Software. *R&D Management, 36*(3), 319–331.

West, J., & O'Mahony, S. (2005). Contrasting community building in sponsored and community founded open source projects. *Proceedings of the 38th Hawaii International Conference on System Sciences,* HICSS 2005, Hawai, USA.

West, J., & O'Mahony, S. (2008). The role of participation architecture in growing sponsored open source communities. *Industry and Innovation, 15*(2), 145–168.

Weston, J. F., Mitchell, M. L., & Mulherin, J. H. (2004). *Takeovers, Restructuring and Corporate Governance.* Upper Saddle River, NJ: Pearson Prentice-Hall.

Wilcox, D. H., Chang, K.-C., & Grover, V. (2001). Valuation of mergers and acquisitions in the telecommunications industry: a study on diversification and firm size. *Information Management, 38*(7), 459–471.

Worek, M., De Massis, A., Wright, M., & Veider, V. (2018). Acquisitions, disclosed goals and firm characteristics: A content analysis of family and nonfamily firms. *Journal of Family Business Strategy, 9*(4), 250–267.

Wu, C.-G., Gerlach, J. H., & Young, C. E. (2007). An Empirical Analysis of Open Source Software Developers' Motivations and Continuance Intentions. *Information & Management, 44*(3), 253–262.

Yin, R. K. (2003). *Case Study Research: Design and Methods (3rd ed.)*. Thousand Oaks, CA: SAGE Publications.

Yin, R. K. (2015). *Qualitative Research from Start to Finish (2nd ed.)*. New York, NY: Guilford Publications.

Yu, Y., Umashankar, N., & Rao, V. R. (2016). Choosing the right target: Relative preferences for resource similarity and complementarity in acquisition choice. *Strategic Management Journal 37*(8), 1808–1825.

Yunker, J. A. (1983). *Integrating Acquisitions: Making Corporate Marriages Work*. Westport, CT: Praeger.

Zaheer, A., Hernandez, E., & Banerjee, S. (2010). Prior alliances with targets and acquisition performance in knowledge-intensive industries. *Organization Science, 21*(5), 1072–1091.

Zaheer, A., Castañer, X., & Souder, D. (2013). Synergy sources, target autonomy, and integration in acquisitions. *Journal of Management, 39*(3), 604–632.

Zeitlyn, D. (2003). Gift Economies in the Development of Open Source Software: Anthropological Reflections. *Research Policy, 32*(7), 1287–1291.

Zhang, X., & Zhu, F. (2011). Group size and incentives to contribute: A natural experiment at Chinese Wikipedia. *American Economic Review, 101*(4), 1601–1615.

Zollo, M., & Singh, H. (2004). Deliberate learning in corporate acquisitions: Post-acquisition strategies and integration capability in U.S. bank mergers. *Strategic Management Journal, 25*(13), 1233–1256.